From This Green Earth

Essays on Looking Outward

Sylvia Engdahl

* *Ad* *
Stellae

Eugene, Oregon
2024

Layout by Sylvia Engdahl

Cover art by Can Stock Photo / ryanking999

Author website: www.sylviaengdhl.com

ISBN: 979-8985853292

Contents

The essays in this book do not need to be read in order. Each essay is independent.

Preface

The original edition of this book, which included many more essays than the current one, was too long to be issued in print form at a reasonable price. I have shortened it in order to offer paperback and audiobook editions of essays about space that I feel are worth preserving in those formats. (The omitted essays and the appendix "Space Quotes to Ponder" will remain available at my website.) It contains one new essay and minor revisions to several others. Because it is more focused on issues of importance than the former edition, even the ebook version may have greater appeal for space advocates, and I hope it will reach a wide audience.

I have been a strong believer in space flight since 1946, and have seen our society's enthusiasm for space rise and fall. Like many others aware of the vital importance of space activity to humankind's future, I have been stricken with despair over the loss of momentum after the moon landings and the failure of the public to recognize the danger of remaining confined to a single planet. My personal interest has been less in space technology than in our need to become a spacefaring species, and for many years the slowness of progress toward that goal both frustrated and frightened me.

But well over a decade ago my view of this lag changed. I was stricken by the analogy between our era's new perception of space and that of the seventeenth century, when the knowledge that the stars are suns surrounded by worlds was first absorbed. It was an emotional shock to the people of that time to realize that they didn't live in a safely-enclosed crystal sphere embedded with lights placed there for human benefit, and it took many years for them to adjust. It dawned on me that the situation today is comparable.

Though for centuries the public has known that the universe is vast, until the first space flights they expected no contact with it. They were not required to imagine what it might mean to leave Earth, or to be vulnerable to visitation from elsewhere. It is not surprising that many shrink from that awareness, subconsciously if not consciously. I now believe that what has seemed to be apathy toward space is in reality suppressed apprehension. In the light of the historical precedent, this is a normal and inevitable reaction that in time will fade.

Because the essays collected here are independent and were written at different times, the book contains considerable repetition of this idea. I feel it cannot be expressed too often, for space supporters tend to be skeptical of it—they have no such underlying feeling themselves and do not grasp its power over the general public. Yet actually, this view of the reluctance to support space activity is a very optimistic one; it reveals that our slow progress in space does not mean that something has gone wrong.

The public has never been behind any major advance in human history. The significant developments have always been made by minorities whose vision was not widely shared. And we see this now that entrepreneurs are at last moving us forward, setting goals that the taxpayers should never have been expected to underwrite. As space advocates, we should rejoice in the knowledge that this is how progress works, and will continue to work in the future as step by step, we move outward toward the stars.

The Once and Future Dream

This is the only new essay in the book, written in 2024. While I have expressed some of these ideas before, it highlights one that I now feel is of particular importance: the impact of the first sight of Earth from Apollo 8. Although it took awhile for the reaction to set in, I believe this, rather than Apollo 11, was the turning point, the day on which, for the second time in history, the public's conception of the universe was forever changed.

*

Once upon a time, more than half a century ago, a dream that had inspired people for over three hundred years reached the first milestone toward its fulfillment. Only an exceptional few had shared it at first, and it had taken a very different form, a form that persisted throughout most of its history as more and more of the educated public became fascinated by the idea. Its modern form emerged in the late nineteenth century. At that time some people began to believe that in the future it might really become possible to travel in space.

The longing to do so, and especially to reach the stars, affected many people from the time they become aware that stars are other suns surrounded by planets. But this was far from the initial reaction to that fact, which had not been even suspected until the Italian philosopher Giordano Bruno suggested it in late sixteenth century.

This was pure intuition on Bruno's part, and because he was not a scientist he has not been given the credit such an innovative concept warrants, at least not in English-speaking countries. It was contrary to the science of his time; even Copernicus, whom he admired, had believed that the stars were mere lights firmly fixed to relatively nearby celestial spheres. It was also contrary to the teaching of the Church, and as a result Bruno was burned at the stake in 1600. Although he was also guilty of other heresies, this was probably the one that caused his inquisitors to imprison him for eight years hoping to make him recant. No doubt they sensed its truth and found it frightening, as did most who heard of it before the late seventeenth century.

Bruno's books were banned, which in itself attracted followers who were excited by them. As the idea spread among more conventional scholars, however, it was extremely upsetting, The thought of stars as huge, fiery spheres at random distances from Earth dismayed people who had believed that the universe was orderly and unchanging, and who had previously associated eternal fire with hell. Moreover, the movement of the planets in our solar system was predictable and had been precisely calculated, which could not be done with stars now conceived as loose in space, As the poet John Donne expressed it, "New philosophy puts all in doubt" and Earth is lost among the stars, "all coherence gone, all just supply and all relation." The philosopher Blaise Pascal's famous statement, "The eternal silence of these infinite spaces terrifies me," may have been the typical reaction of his contemporaries. A universe that isn't neatly arranged and run like clockwork does not feel safe.

It took a long time for this feeling to die out. Nevertheless, the new cosmology took hold as scientific support for it grew, beginning in 1610 when Galileo published an account of his observations of the sky through a telescope. These showed that contrary to previous belief, heavenly bodies are not perfect, unblemished spheres; worlds other than Earth exist, if not other suns. Later a theory of Rene Descartes about fluid vortices carrying them through space overcame objections based on traditional physics, which held that there could be no such thing as a vacuum. These developments led the way for the very first suggestions of future space travel and even of space colonization.

In a letter written to Galileo in 1610 astronomer Johannes Kepler wrote. "As soon as somebody demonstrates the art of flying, settlers from our species of man will not be lacking. . . . Given ships or sails adapted to the breezes of heaven, there will be those who will not shrink from even that vast expanse." It is unlikely that anyone else saw this private letter, but in 1638 John Wilkins, a bishop of the Church of England, published a long and very influential book about the moon in which he declared, "It is possible for some of our posterity to find out a conveyance to this other world, and if there be inhabitants there, to have commerce with them."

There is no evidence that this was taken seriously by his readers, for no similar comments followed. Although stories about voyages to the moon appeared during the next two centuries, they were meant as pure fantasy. Interest in actually seeing other worlds took another path entirely. But first, the public learned of the existence of multiple suns with planets through the first popular book about them intended for laymen rather than scholars, *Conversations on the Plurality of Worlds* by French author Bernard de Fontenelle. It became a bestseller, perhaps because unlike all other science books of that time it was specifically directed to women. Fashionable ladies were entranced by it, and it forever transformed the idea of multiple worlds from a dreadful thought to a fascinating one.

Two major developments firmly established the new cosmology as the standard view of the universe by which few people were disturbed. First, an assumption about other worlds' purpose, emphasized by Bishop Wilkins and considered conclusive for the next two and a half centuries, caused religious authorities to stop opposing the idea of other worlds and become its most ardent defenders. Virtually everyone agreed that God could not have created a world without any purpose. Therefore it must be of use to someone, and since distant planets were of no use to Earth (a presumption today's space colonization advocates might question) it must have inhabitants of one kind or another. A picture emerged of countless inhabited worlds, attesting to the power and glory of God.

The other major factor in the triumph of the new cosmology was Isaac Newton's discovery of gravitation. Initially the idea of gravity was rejected by science because it was seen as an "occult" force, much as conventional scientists of today view ESP. All authorities insisted that nothing short of divine intervention could act at a distance. But Newton's mathematical laws of planetary motion explained things that could not be explained in any other way. Moreover, they restored confidence that there is order in the universe—people no longer had to face the frightening thought of planets drifting without any pattern. So gradually the Newtonian system prevailed.

By this time educated people who read current books and

magazines had heard a lot about the new astronomy and most were eager to know more. There was much speculation about other planets' inhabitants, who were almost always assumed to be superior to humankind. Some thought they were angels, while others viewed them as mortals without the flaws all too evident in natives of Earth. A great deal of poetry, some of it book-length, presented detailed descriptions of spectacularly beautiful suns and their attendant worlds. Often it took the form of space voyages, not science fiction but imaginary journeys presented as having ocurre4 during a dream or with a spirit guide.

Enthusiasts found it very frustrating to think that they could never really see distant regions of space—that as far as anyone knew, no humans ever could. And so some began to dream of the future, not posterity's future but their own. Though it was impossible for mortals to travel in space, they thought they might do it after death on their way to Heaven.

A British clergyman, William Derham, wrote a very popular book titled Astro-Theology. It was mostly about astronomy, but at the end of it he said, "We are naturally pleased with new things; we take great pains, undergo dangerous voyages, to view other countries: with great delight we hear of new discoveries in the Heavens, and view those glorious bodies with great pleasure thro' our glasses. With what pleasure then shall happy departed souls survey the most distant regions of the universe, and view all those glorious globes thereof, and their noble appendages with a nearer view?"

Though this was not the doctrine of any church, it was not incompatible with existing concepts of an afterlife and many clergymen espoused it. The idea caught on, or perhaps was intuitively felt by people who simply couldn't bear the thought of never seeing more of the universe than was visible from Earth. Richard Gambol's long poem Beauties of the Universe sums it up, telling of a time when the soul

> Unbounded in its ken, from prison free
> Will clearly view what here we darkly see:
> Those planetary worlds, and thousands more,
> Now veil'd from human sight, it shall explore.

Shortly after the death of Newton, who was revered as the greatest scientist who had ever lived, there were many such poems; people couldn't believe that so great a man wouldn't have a chance to see the worlds whose motion he had explained. As one woman put it, "With faculties enlarg'd, he's gone to prove / The laws and motions of yon worlds above; / And the vast circuits of th'expanse survey, / View solar systems in the Milky Way."

David Mallet's *The Excursion: a Poem in Two Books* devotes many pages to what Newton's soul perceives:

> Ten thousand worlds blaze forth; each with his train
> Of peopled worlds. . . But how shall mortal wing
> Attempt this blue profundity of Heaven,
> Unfathomable, endless of extent!
> Where unknown suns to unknown systems rise,
> Whose numbers who shall tell? stupendous host!
> Sun beyond sun, and world to world unseen,
> Measureless distance, unconceiv'd by thought!
> . . . What search shall find
> Their times and seasons! their appointed laws,
> Peculiar! their inhabitants of life,
> And of intelligence . , , each resembling each,
> Yet all diverse! . . .
> About me on each hand new wonders rise
> In long succession.

In prose, too, descriptions of souls' travel to the stars appeared during the eighteenth and nineteenth centuries. In a popular book published in 1840 Thomas Dick wrote, "The law of gravitation . . . separates man from his kindred spirits in other planets, and interposes an impassable barrier to his excursions to distant regions, and to his correspondence with other orders of intellectual beings. But in the present state he is only in the infancy of his being. . . He will, doubtless, be brought into contact and correspondence with numerous orders of kindred beings, with whom he may be permitted to associate on terms of equality and of endearing friendship. . . It may be necessary that branches of he universal family that have existed in different periods of duration, and in regions widely separated from each other, should be brought into mutual association, that they may communicate to each other the results of their knowledge and experience." That looks like an astonishingly modern prophecy until seen in context, which shows it to refer to the afterlife. People's hopes and aspirations don't change; they merely take the form appropriate to the era culture in which they are expressed.

Today's readers are apt to dismiss these writings as merely religious, without relevance to the growth in human understanding of the actual universe. But there was no firm line between religion and science in that era. Only more recently were they separated by more than the availability of data to analyze; the word "scientist" was not coined until 1840 and has been retroactively applied to the few who had relevant facts to study. All unrelated speculation was envisioned, even by them, in the only terms people knew—no doubt metaphorically by some although the majority took such speculation literally. There was no data about exoplanets and no imaginable means of acquiring any, so naturally people fit thoughts about them into their existing framework of belief.

Their version of the dream of going to the stars was not totally abandoned; it simply took different forms as the views of the majority changed. When the concept of Heaven as a physical place faded, some people came to believe that worlds of distant stars are the happy abodes of the dead, or more recently, that souls are successively reincarnated on different planets. The latter belief is quite common among adherents of New Age ideas even today.

Though interstellar travel may eventually become possible, no one alive now will ever experience it. Yet the longing is as strong as ever, and many would understand, if not share, what nineteenth-century poet Henry Abbott felt when he wrote, "Seek how we may, / There is no other road across the sky. / And, looking up, I hear star-voices say: / You could not reach us if you did not die."

*

The modern version of hope for space travel was born in 1865 with the publication of Jules Verne's novel *From the Earth to the Moon*, which for the first time suggested that technology might make it possible. In spite of the opinions of certain narrow-minded people, who would shut up the human race upon this globe, as within some magic circle which it must never outstep, he wrote, we shall one day travel to the moon, the planets, and the stars, with the same facility, rapidity, and certainty as we now make the voyage from Liverpool to New York This remark is even more relevant now than when he made it.

Verne may or may not have been serious about interstellar travel, but Winwood Reade was in his 1872 bestseller *The Martyrdom of Man*, a long, controversial history of humankind that may be the first mention of it in nonfiction. He wrote, A time will come when science will transform [our bodies] by means which we cannot conjecture... And then, the earth being small, mankind will migrate into space, and will cross the airless Saharas which separate planet from planet, and sun from sun. The earth will become a Holy Land which will be visited by pilgrims from all quarters of the universe. Although the book was widely read, little if any notice was taken of this paragraph. which was no doubt considered too radical to be viewed as a prediction.

During the last few decades of the nineteenth century a number of novels involving space travel appeared. In contrast to past centuries' imaginary flights to idyllic worlds, most featured thrilling adventure, as science fiction has ever since, and the dream of going into space became a focus of the human need for excitement and challenge. Also during this period, hope of actually contacting another world arose, for the sighting of features on Mars thought to be canals convinced the public that there were technologically-advanced Martians and a number of proposals for sending signals to them were made.

In 1898 H. G. Wells' novel *The War of the Worlds* made the first suggestion of a potential dark side to space travel. No one before him had imagined that inhabitants of other planets might not be friendly. The then-current hope of contacting real Martians precluded any effect the novel might have had on feelings about space. Not until it was dramatized in the form of news in a 1938 radio broadcast was there panic, and that was shirt-lived once the public learned that the news had been fiction The time was not yet ripe for alien invasion drama to reflect underlying emotion. But the idea of hostile visitors had been planted in the collective unconscious, to be enhanced by the countless stories about interstellar war that have appeared since.

The Russian rocket pioneer Konstatin Tsiolkovsky wrote extensively about space in the late nineteenth and early twentieth centuries, including philosophical works less well-known than his technical ones. He described orbiting habitats in detail and believed that humans will eventually colonize the galaxy. They will control the climate and the solar system just as they control the Earth, he wrote. They will travel beyond the limits of our planetary system; they will reach other Suns and use their fresh energy instead of the energy of their dying luminary. But while his work had a strong influence on the Soviet space program, most of his other ideas about humanity conflict with Western views and few of his non-technical writings were translated. Their chief global significance lies in demonstrating that people of all political and religious persuasions share the dream of going to the stars.

During the 1920s and 1930s actual work on rocket technology was carried on by a small number of visionaries, though as recently as 1936 an article in *Scientific American* declared that it would never be possible to go to the moon because while rockets might power a spaceship there would be no way to navigate (the advent of computers was not foreseen). That didn't slow the growth of public interest in science fiction, which was read by a relatively small number of devotees in the twenties but spread in the thirties and forties to a much wider audience. The comic strips Buck Rogers and Flash Gordon appeared in Sunday newspapers read by millions. Children whose parents had grown up on Westerns were now addicts of their radio and movie serial versions, and in the fifties of the television *Space Cadet*. The pulp science fiction scorned in literary circles was supplemented by work of higher quality, some of it in mainstream publications.

Why did this happen at this particular time? It was largely because the technological developments of the early twentieth century—automobiles, electricity in homes, and especially aviation—made people expect more innovations in the future. And to many the future was symbolized by idea of other worlds. The dream of traveling beyond Earth had fascinated people for nearly three hundred years. Now, enhanced by the prospect of rockets, it spread through the collective unconscious of living generations. In 1946 when I was twelve years old, the teacher in a science class read aloud a description of what it would be like to travel in space. I knew at once that this was true and important, though I had never been exposed to any science fiction and was totally ignorant of relevant technology. I told a friend a few days later that I was sure people would get to moon within twenty-five years, an amazingly accurate guess considering that I had no basis for it. I simply sensed that it was time we got to the moon, and evidently many others did, too.

Not all thoughts about space were positive. The V2 rockets of World War II created an underlying suspicion that danger might someday come from beyond the sky, which was expressed by the alien invasion movies of the 1950s, Unlike more recent ones these movies made no attempt at realism; even to audiences of the time they caused the notion of invaders from space to seem ridiculous, which was the root of their popularity. Though the science of the time rejected the assumption that other planets are inhabited, generations of past belief had established the idea deep in the collective unconscious, and people were no longer so naive as to think that the inhabitants are like angels. Thus this was also the era of the first UFO reports, People who believe in UFOs are not crazy; they simply perceive concepts metaphorically rather than with the logical portion of their minds. Their accounts of what they have seen reveal much about a culture's suppressed thoughts about aliens.

The launch of Sputnik in 1957 changed human outlook on space emotionally as well as in terms of the impact on technology. For the first time, activity beyond this planet was viewed as a part of real life. Initially some people were upset by fear of Soviet superiority, but once it became evident that America could catch up in space they shared the general feeling of excitement. Many aspects of popular culture adopted a space theme, which was often used in advertising because the response was generally positive. Though there were opponents of the space program, their objections were based merely on their belief that the money it cost could be better spent on something else. And so as success followed success, from the first man in orbit to the first spacewalk and on to the effort to actually reach the moon, progress in space became an anchor of optimism in the otherwise troubled and turbulent sixties.

July 20,1869, when the Apollo 11 astronauts landed on the moon, was as *Life* magazine called it, "the day man left his planetary cradle." An estimated 650 million people worldwide, from Paris and Rome to rural Africa and Lapland, watched on television and cheered.as Neil Armstrong took his first step on the lunar surface and declared, "One giant leap for mankind." Walter Chronkite, the most prominent news anchor of the era, said, "Man's dream and a nation's pledge have now been fulfilled. The lunar age has begun. And with it, mankind's march outward into that endless sky from this small planet circling an insignificant star in a minor solar system on the fringe of a seemingly infinite universe." G. Crowther wrote in the magazine *New Scientist* that it "can be compared with the original emergence of life from the primordial ocean; it sets a new stage for evolution," a view then shared by many.

And yet the excitement did not last. Despite the continuing enthusiasm of a minority, the public at large turned away from space. Why such a sudden reversal? It is generally said that people had cared only about the space race between the United States and the Soviet Union, and that once it was won they lost interest, This may have been true of some and it was certainly the case with respect to government funding of the space program, but it cannot account for the abrupt loss of feelings that had pervaded Western culture for nearly three hundred years. During most of that time competition in space was unheard of, yet people would have been thrilled to see spaceships explore planets beyond Earth. Those of the past who viewed the sight of them as the hoped-for reward of the heaven-bound after death would have found abandonment of such exploration inconceivable, as would those who were fascinated by the prospect throughout the period when it was seen as a future goal of technology. Nobody ever imagined that humans would reach the moon and then just stop. In the immediate aftermath of Apollo 11 people expected that a trip to Mars would follow before the end of the century. What made so many lose sight of that hope?

In historical accounts of the climactic Apollo 11 moon landing, the importance of the preceding mission Apollo 8, which orbited the moon on Christmas Eve 1968, is often overlooked. Yet actually, it was in several ways more significant. To leave Earth for the very first time and arrive at another world was no small achievement. Apart from the daring it involved, it had a greater impact on human consciousness than is generally recognized. As the astronauts sent their Christmas message, reading from the Book of Genesis while millions watched on television, some people felt as those of the seventeenth and eighteenth centuries would have, awed by the new world because of their belief that God created I, while others were struck by the continuity of human culture as shown by the use of an ancient text. But underneath, most were shaken by the realization, later enhanced by Apollo 11, that unlike the worlds of their dreams the moon—which traditionally had romantic connotations—is very, very dead. Thus more momentous than the immediate reaction was Apollo 8's transformation of the public's attitude toward our home world.

No one had left Earth before. No one had seen it whole, stunningly blue in contrast to the blackness of space. That sight, and the color photo Earthrise taken of it, had an unexpected effect on astronauts and the public alike. They aroused feelings for Earth that hadn't existed when the focus had been on exploring what lay beyond. After the dismaying deadness of the moon, it was suddenly perceived not only as uniquely beautiful but as small and somehow fragile, in need of protection from harm. And so, for good or for ill, the environmental movement was born. `

The good lies in the realization that preservation of Earth's beauty is important and that it does need to be protected from wanton destruction by humans who give no thought to the future. It is less rational to assert that no other spot in the universe is worth anything and to maintain, as many environmentalists do, that humankind can never thrive elsewhere. But it's not a matter of reason, any more than is the reluctance to support space exploration on the part of people who seemingly just don't care. Such reactions are driven by emotions of which these people are often unaware—the same underlying emotions that caused seventeenth-century people to be slow in accepting the new cosmology of their era.

It's not surprising that these emotions have resurfaced. Before Apollo the idea of space travel was abstract—fascinating to think about, but not of real concern to anyone not directly involved. It was thrilling to watch brave men walk on the surface of the moon, and natural to take pride what humans had accomplished. But to be suddenly stricken with awareness that it was all *real*, that Earth is one very small planet afloat in the vastness of space, vulnerable to whatever perils may appear out of nowhere, was a serious shock. How could people not react with apprehension? How could they not tell themselves that it would be unwise or even unethical to pursue space travel when there are enough problems to deal with at home?

The Irony of it is that the only solution to most of those problems involves going into space and making use of its resources—and that in fact, if an asteroid should approach, without a space-based defense the supposed fragility of Earth would become all too real. But people do not want to know that. When it's explained by space advocates they close their minds. Little more than a year after Apollo 11's success the last three planned moon missions were cancelled for lack of support. Though a limited amount of near-Earth orbital activity was undertaken and robotic landers were sent to Mars, space had ceased be a compelling topic of public interest. An insulating layer of boredom had dampened the enthusiasm people once had for it Eventually a few went so far as to convince themselves that the moon landings didn't really happen.

The desire to see, or at least imagine, worlds unlike our own has not faded, but its focus has changed. Now for entertainment people turn to fantasy, much it based on themes or legends with comforting ties to tradition. Though there is still an audience for science fiction, movies about space, with the exception of Star Trek, have become more and more negative. The ultimate expression of this trend was the conclusion of the 2004-2009 version of *Battlestar Galactica*, in which the protagonists, after long years in space, destroy their technology by sending their starships into the sun and retreat to a low-tech lifestyle on prehistoric Earth. Nobody on the Internet seemed to view this as a tragedy.

Yet this does not mean we are confined to our ancestral world forever. Like a child who puts one toe into cold water and pulls back, we sampled the new medium and many shrank from it, But the hesitancy to plunge will not last. Just as to seventeenth-century people a universe full of innumerable uncharted suns did not feel safe, so in our era one in which Earth appears to be the sole haven in a vast expanse of darkness does not. But there are still dreamers—because the population is larger and the need greater, there are more of them than in the past. Activists and entrepreneurs, they are leading the way to a future in which humankind will spread far beyond one small and isolated planet. Fifty-three years passed between the banning of Galileo's book about astronomy and the publication of Fontenelle's hugely popular one. It has now been fifty-four years since the cancellation of the last planned Apollo missions. We are due for a shift in public response.

Sooner or later there will be another new perception of the universe that people shrink from. Strange dimensions? Alien civilizations? Who knows? Yet as always, there will be dreamers to carry humankind forward toward contact with the unknown. The dream of reaching the stars is timeless. However challenging the future may be, that dream will prevail.

Thoughts on the 50th Anniversary of the First Moon Landing

I posted this essay at my website on July 20, 2019. It is more optimistic than anything I could have written ten years previously—progress, slow as it may seem, does occur.

*

Half a century ago it was widely recognized that the first moon landing marked a new era, an era when, having taken the first step toward becoming a spacefaring species, we would never again be confined to one small planet. And yet, to the dismay of a good many of us, our confidence proved to have been premature. After five more moon landings in close succession we haven't ventured beyond low Earth orbit. What went wrong?

I believe nothing went wrong—our only mistake was in our expectations. Apollo 11 was, and always will be, an unsurpassable milestone in the history of humankind. However, it set precedents that left us with false assumptions: not only an overly-optimistic idea of what would soon follow, but miscon

ceptions about what it would take to establish an expanding human presence in space. Apollo 11's success was the result of a necessarily huge and expensive tax-funded project, which could not have been launched without public support. And so we assumed that future advances would always be achieved by large tax-funded projects and that public support would therefore always be needed, and could be obtained. But human progress does not ordinarily happen in that way.

If Columbus and other explorers who sought to reach new lands had needed support from the public, they could never have set sail. The settlement of those lands, though chartered and sometimes initially financed by European governments, was expected to be profitable and did not depend on the approval of the population left behind. By definition, pioneers in any field of endeavor are people with exceptional vision that is not shared by the majority of their contemporaries. It is no coincidence that we're at last moving forward in space now that private companies are taking over, something that would not have been possible without a long period of low-orbit preparation. Nor is it surprising that in regard to ambitious plans for larger projects made by NASA, the availability of sufficient funding is questionable.

Yet the Apollo program did get funded. The public did support it. Thus space advocates expected that this would continue and were puzzled and disappointed when it did not. It's generally said that the public's real enthusiasm had been for winning the Cold War and that once we had beaten the Russians to the moon people lost interest; and as far as the appropriation of funds by Congress is concerned, that was probably true. But I think it's a good deal more complicated than that.

Twenty-five years ago—halfway into the hiatus that's now ending—I taught an online Media Studies course on the Mythology of the Space Age [*]. At that time there was still strong interest in space if not in the provision of funds, and it was evident that a new mythology comparable in significance to ancient mythologies was emerging—which to me was a very hopeful sign. I focused the course on the optimistic view expressed in pop-culture space fiction, in contrast to the gloom that pervaded portrayals of future Earth. Since then, however, the darker side of Space Age mythology has become more prominent while the bright side has diminished. There are, of course, a great many people whose desire to see humans explore beyond this planet has not weakened. But a smaller percentage of the population now welcomes the thought that we are part of a vast and mysterious universe.

One form of evidence for this is the increase in pessimistic space movies, as opposed to the primarily upbeat ones of the seventies and eighties. Another is the widespread exaggeration of environmental concerns, especially the illusion that Earth can remain unchanged from generation to generation and that we can keep it that way, even to the extent of controlling its climate. And still another is the decline of public support for real space programs, which has frustrated space advocates for nearly half a century—contrary to common belief, that decline wasn't due to boredom or apathy. Until recently I did not grasp the ambiguous nature of the mythology that was developing. I now feel that many people's enjoyment of space fantasy was a matter of whistling in the dark.

According to one anthropological theory of mythogenesis, the one I personally favor, mythology serves the adaptive purpose of helping people confront the environment in which they perceive their culture to exist. If the perception of the extended environment changes, mythology not only changes but goes through a period of upheaval. When the first photos of Earth as a globe made clear that our world, not just in science textbooks but in reality, exists within a seemingly-endless expanse of empty space, they produced an emotional shock. In part, this was good; people became aware that Earth is beautiful and should be protected from wanton destruction. And some were elated by the possibility of venturing into new regions.

But other people withdrew, telling themselves either that nothing beyond our planet matters, on one hand, or on the other, that space travel was merely a fascinating new game to watch. Then came the moon landing, after which it could be dismissed as a game no longer (although some have tried by claiming that the landing never happened). Though it took awhile for this reaction to set in, the eventual result was a shift in mythic views of the future. At first, people inwardly disturbed by what Pascal famously described as "the eternal silence of these infinite spaces" had nevertheless been drawn to space and had joined the enthusiasts in looking at it as a pleasantly exciting setting for stories; thus optimistic science fiction movies were popular while at the same time, the moon was abandoned. Until the *Challenger* disaster, which brought to the surface subconscious feelings that had been building up since Apollo, the potential terrors of the unknown were deemphasized; and the idea that if we could travel in space then visitors from elsewhere might also do so was dealt with by everyone but UFO believers through the portrayal of such aliens as friendly.

Gradually, however, this comforting outlook faded. Aliens are now depicted as dangerous; some scientists have begun to say that we shouldn't attract their attention by sending SETI messages. Space activity arouses less interest than it once did, even among children, and the majority of citizens bury their heads in the sand with regard to the fact that a single planet's resources cannot possibly support its population forever. For escapist entertainment, most now look to pure fantasy about magical realms rather than wishing "to explore strange new worlds, to seek out new life and new civilizations, to boldly go where no one has gone before." Unconsciously if not consciously, they would prefer not to think about what lies beyond our home world.

I have come to realize that we should have never have expected it to be otherwise. For the vast majority of humankind, apprehension always has been the underlying reaction to new awareness of our place in the universe. In my essay "Confronting the Universe in the Twenty-First Century" (available at The Space Review, at my website, and in this book as well as in my book The Planet-Girded Suns) I compared our present situation to the seventeenth-century spread of knowledge that Earth was not the center of the universe and the stars were not mere lights fixed to a surrounding crystal sphere. That took people a very long time to absorb. And though our forebears did absorb it and became fascinated by the idea of inhabited worlds orbiting other suns, they did not connect that idea with everyday reality. They didn't expect those worlds to have any impact on the inhabitants of this one, let alone imagine that someday mortals might leave it.

Not so today, when for better or for worse, the prospect of contact with the wider universe is before us. And I'm sure that ultimately Space Age mythology will attenuate the terror of it and enable our culture as a whole to meet it with confidence—the metaphors of myth are, after all, the means by which humans deal with the unknown. I cannot guess when this will happen, but I now see that it need not be soon. Though it is not possible to convince today's taxpayers of the importance of space, it is not necessary, either.

Movement outward from this planet is not as dependent on public support as I and other space advocates once thought; entrepreneurs are beginning to take the initiative, as has happened in every other area of ongoing progress. In all probability the first ship to Mars will be launched not by the government but by Elon Musk. This is how human advance works. Of necessity, it depends on individual vision, for instinctively people sense that survival of our species (or any species) depends on new ventures being made by a few before the entire population can safely commit to them.

Thus new conceptions of reality are always embraced by a farsighted minority before being widely accepted, and the key to acceptance is not persuasion, but time. The fact that fifty years of marking time have passed since humans first set foot on the moon is cause not for pessimism, but for gladness that so much of the time is now behind us. We are on track, and in due course, when travel to the moon and Mars becomes routine, mythic expressions of our place in the universe will again serve to inspire the widespread feeling that this is how the future will be, and was always meant to `be.

* "Science Fiction and Space Age Mythology," a Media Studies course offered in 1994 and 1995 by Connected Education, a pre-Internet online conferencing program directed by Paul Levinson, for graduate credit from the New School for Social Research. The "lectures" for it are online as "The New Mythology of the Space Age."

Breaking Out from Earth's Shell

Long ago people literally believed that an invisible, transparent shell surrounded earth to hold up the stars. after that theory was discarded, the shell remained in a figurative sense, for earth was thought to be forever isolated from the rest of the universe. some people are comforted even today by this illusion, although it's time for humankind to break free. This essay also appears at the Lifeboat Foundation's website.

*

Since the dawn of history humans have been drawn to the idea of reaching the heavens. Most cultures' gods were presumed to dwell there. The stars have always symbolized mankind's highest aspirations. They were, however, viewed as inaccessible, except perhaps by souls after death.

According to the worldview universally accepted in Europe prior to the seventeenth century, the realm of the fixed stars (as distinguished from planets) was literally perfect and unchanging. The stars were thought to be embedded in an invisible sphere—a shell—that encircled Earth, for if they were not attached to something, would they not fall? Contrary to a common misconception, the theory that Earth is not the center of the solar system was not resisted because of any sense of demotion from the place of prime importance, or even because of conflict with the Bible—actually, it would not have aroused much opposition if the authorities of the time had not realized, more than fifty years after its publication, that it opened the door to questioning the nature of the stars. Copernicus himself never doubted the accepted theory; to him, they were still firmly attached to a crystal sphere surrounding the sun.

However, near the end of the sixteenth century the philosopher Giordano Bruno, a strong advocate of the Copernican theory, suggested for the first time that the stars are not mere lights in the sky, but suns with planets of their own. Although that was not his only heresy, many scholars believe it was the primary reason why he was burned at the stake, and why, after his books were banned, supporters of his astronomical ideas called themselves Copernicans instead of mentioning his name. Rearrangement of the solar system was a relatively minor issue compared to the upheaval in both science and religion caused by denial of the stars' perfection.

To people who had believed themselves safely enclosed within a perfect sphere, beyond which lay Heaven, the idea of a universe full of suns at random distance from Earth was extremely upsetting. As John Donne put it in these famous lines from his poem "An Anatomy of the World" (1611), it removed all coherence from their worldview.

> And new philosophy calls all in doubt,
> The element of fire is quite put out,
> The sun is lost, and th'earth, and no man's wit
> Can well direct him where to look for it.
> And freely men confess that this world's spent,
> When in the planets and the firmament
> They seek so many new; they see that this
> Is crumbled out again to his atomies.
> 'Tis all in pieces, all coherence gone,
> All just supply, and all relation.

This was not a foolish or naive reaction. Human thought is dependent on a stable foundation on which to rely. Psychologically, people are cast adrift if their basic premises are questioned, and this is an adaptive trait since without anchors it would be impossible for a society to function. Change must come gradually, through the exceptional few who are able to discard the outlook of their contemporaries. Once they do, a new outlook spreads—but that takes time.

Over time, people became reconciled to the loss of an immutable order in the universe, and by the late seventeenth century those with enough education to care about astronomy envisioned countless suns, all surrounded by planets which, like those of our own solar system, were assumed to be inhabited. Some thought they were the homes of angels or the souls of the dead, but the belief that they were inhabited by mortals superior to ourselves soon predominated. The one thing everyone agreed about was that they were not without tenants. It had formerly been believed that the heavenly bodies existed for the benefit of humankind, so since distant suns and planets were of no apparent benefit to us, it was reasoned that they must have been made for other mortals. It was taken for granted that God would not have created a "useless" world. This was not questioned until the middle of the nineteenth century, when after hot debate the conviction that all extrasolar worlds have inhabitants still prevailed. Not until early in the twentieth was it abandoned.

Ironically, we now know that uninhabited planets are not useless—it may well become possible for us to colonize them. They may prove to be our salvation when the resources of Earth are gone. But of course in earlier centuries that possibility did not occur to anyone.

For more than three hundred years it was believed, by educated people at least, that solar systems similar to ours exist. This is known because they are referred to in the writings not just of astronomers but of many well-known people such as Benjamin Franklin as well as in the popular magazines of the day, sermons, and even textbooks for children. Also, a great deal of poetry, some of it book-length, was written about spectacular suns and their planets. Imaginary voyages through space appear even in poems by major poets such as Milton, Shelley, and Byron.

These were spiritual voyages, not journeys in spaceships. For example, at the time of Newton's death it was often suggested that he might see distant planets on his way to Heaven. Many people longed to visit the worlds so frequently talked about, and doing so in an afterlife was the only route they could imagine. As late as the 1870s the American poet Henry Abbey wrote:

Death, that dread annulment which life shuns,
Or fain would shun, becomes to life the way,
The thoroughfare to greater worlds on high,
The bridge from star to star. Seek how we may,
There is no other road across the sky;
And, looking up, I hear star-voices say:
"You could not reach us if you did not die."•

But the longing for a closer look at other worlds was not shared by everyone. Searches for information about it turn up only what was written by those who were interested in cosmic space. Many who heard of distant solar systems were not interested, or may even have been disturbed by the thought. The French philosopher Pascal famously wrote, "The eternal silence of these infinite spaces terrifies me," and he can hardly have been alone in feeling that way. It was not an issue people needed to be concerned about. Earth was, after all, safely isolated from the larger universe, as far as they knew. No one supposed that there could ever be actual contact could between worlds; it was as if an invisible shell still enclosed our own.

It is likely that the poet William Wordsworth's feeling about space was—and still is—more typical than that of the space enthusiasts. He was knowledgeable about astronomy and enjoyed watching the stars with his sister and friends. But he is best known for his love of nature. When in his famous poem "Tintern Abbey" (1798) he wrote:

> Therefore am I still
> A lover of the meadows and the woods,
> And mountains; and of all that we behold
> From this green earth

was he considering the stars a part of nature? There is no indication in the poem that he was, yet it seems unlikely that he would have used the phrase "*from* this green earth" rather than the more common "*on* this green earth" if he had never looked outward, thinking of Earth as part of the larger natural universe. And in fact in another poem, "The Stars are Mansions Built by Nature's Hand," he viewed worlds of the stars as the happy abode of the dead.

Be that as it may, it was Earth alone that he cared about. In "Peter Bell," describing his return from a fantasy space journey within our solar system, Wordsworth revealed an outlook that is shared by many today, more than two hundred years later.

Swift Mercury resounds with mirth,
Great Jove is full of stately bowers;
But these, and all that they contain,
What are they to that tiny grain,
That darling speck of ours!

Then back to Earth, the dear green Earth:
Whole ages if I here should roam,
The world for my remarks and me
Would not a whit the better be;
I've left my heart at home.

See! there she is, the matchless Earth!
There spreads the fam'd Pacific Ocean!
Old Andes thrusts yon craggy spear
Through the grey clouds—the Alps are here
Like waters in commotion.

And see the town where I was born!
Around those happy fields we span
In boyish gambols—I was lost
Where I have been, but on this coast
I feel I am a man.

The last few lines say it all: for the vast majority of people, their very identity depends on their presence on Earth. To leave it in fantasy is one thing, but to be aware that people can really leave, really venture into unknown regions, puts a whole new face on facts that have been known for centuries. And when astronauts do leave, even they are often more deeply moved by the sight of "the dear green earth" from a distance than by the beauty of the stars. While astronauts are exceptional individuals who enjoy space flight and long to explore the universe, those are not the emotions the public vicariously shares.

The farther we go in space, the further removed space travel is from theory, the more evident this will become. In Wordsworth's time and for two centuries longer, the idea that there might be peril in space didn't occur to people. Their knowledge of the universe was very abstract. Even the few brief mentions of traveling to the stars that appeared in the late nineteenth century did not suggest that it would be dangerous, and certainly there was no suspicion that the presumed inhabitants of other worlds might not be friendly. That notion was introduced by H. G. Wells' 1897 novel *The War of the Worlds*, which was viewed as pure fantasy until a 1938 radio dramatization was broadcast in the form of a news report, nearly causing a panic. Science fiction of the 1920s was read only by those especially interested in ideas about the future. However, starting in the 1930s, the hit comic strip and radio adventures of Buck Rogers forever changed the public's perception of space. It became the scene of violent action and exciting new concepts, and the development of V-2 rockets in World War II led to a suspicion that there might possibly be some truth in them.

Thus the first UFO sightings, which occurred in 1947 and were immediately associated with extraterrestrials, were followed by countless alien invasion movies in the 1950s. It is often said that these movies were actually about the Cold War, and no doubt their plots were influenced by it; but I believe that underneath, they reflected the public's new realization that space may hold terrors. These films featured ridiculously-portrayed aliens and some were intentionally humorous, which suggests that viewers wanted to think that the whole idea of danger from beyond Earth was silly. When the development of satellite technology began to show that space travel is not silly, their popularity waned. People turned their attention to the competition in space, which really was connected to the Cold War. After America won they could no longer be distracted by it, and the worries suppressed so long began to surface along with growing anxiety about our own ventures into the unknown.

At the time of the Challenger disaster I was astonished by the widespread public feeling that it meant space travel shouldn't be undertaken, and especially that the civilian teacher Christa McAuliffe shouldn't have been "sent" into space, as if she hadn't been chosen out of thousands of applicants who vied for the chance to go. The supposition, sometimes even explicitly stated, that she hadn't known it was dangerous was an insult to her courage and to her intelligence. Who could possibly be unaware that riding in a spacecraft propelled by rocket engines and boosters providing 7.8 million pounds of thrust at liftoff involves risk?

Perhaps previous space flights had been viewed with a sense of detachment, as if they were science fiction. But I now think there was more to it than that. It would simply not have been rational for anyone ever to think space travel isn't dangerous; the evidence that it's unsafe could hardly have come as a surprise. And even if it did, dangers involving far greater numbers of people, such as those of early aviation, had been accepted by the public without question. No one said that pilots shouldn't be allowed to take off in primitive planes, although the crash rate was extremely high. Planes, however, did not get very far from the ground. There was no possibility that improved ones would leave the planet and enter unknown regions beyond. I suspect that the realization that space travel is *real* came not with the tragedy of *Challenger*, but with the Apollo moon flights, and that *Challenger* brought to the surface unconscious feelings that had been building up for a long time. Underneath, people were troubled not by the danger to the astronauts but by the potential perils of contact with the wider universe. The Challenger disaster was merely the trigger for expression of the public's growing uneasiness about spacefaring.

As in the seventeenth century, for people to shrink from the necessity of revising their perception of our environment (our total environment, not the mere biosphere) is normal and adaptive for our species. If everyone's orientation shifted suddenly, society would disintegrate. Civilization depends on the ability to make plans knowing that for the short term, tomorrow will be like today. Thus changes in outlook come slowly, first in a few visionaries, later in one generation after another as minds open to new awareness. Mass media, a recent phenomenon, will speed up the process but cannot make it happen overnight. If anything, real-time mass media events such as the moon landings produce more shock than lasting transformation.

So it can't be expected that the public will be quick to support future space activity. We should not be surprised if more interest is shown in science fiction movies than in real flights. Inwardly drawn to the thought of venturing beyond Earth but unable to break away from the safe and familiar, people tell themselves that the fiction they enjoy is mere entertainment, not be taken seriously. Though science fiction is sometimes criticized for promoting unrealistic dreams, more likely it serves to assure the average moviegoer that they are *just* dreams, while inspiring the visionary minority to work toward making them come true.

Will we always be bound to Earth, then? Of course not. Evolution is slow, but it can't be halted. Humans have been seeking new lands to settle for millennia, first new villages, and eventually new continents. The negative expression of this instinct has been the urge to expand a group's territory through war, which hopefully most of us have outgrown since the times when young men dreamed of glorious conquest. On the positive side, there has always been a desire of ordinary people to be pioneers, even at the cost of hardship. For some time it has been evident that Earth has, or soon will, run out of vacant land. How could there not be an impulse to go beyond, quite apart from the plain fact that a species that fails to move beyond the niche in which it has evolved must be periodically decimated or else become extinct? In the long run, how could humankind fail to follow that impulse?

The dream of extending our species' range beyond the world on which we evolved is hardly something so trivial as entertainment, however much entertainment may be derived from it or how gradually it is absorbed. It is an often-unconscious expression of the deep-seated instinct present in all species to expand their ecological niche, an adaptive response to the ever-present threat of extinction. It has become trite and unfashionable to compare movement beyond Earth to the movement of life from the sea onto land, as was done during the Apollo era, but that comparison is still valid.

These are not new ideas—space advocates have been expressing them for years. Why then have so many lost sight of them and become discouraged? I think it is because in our era people are so used to rapid change, and to instant gratification of their wishes, that they have lost all sense of the evolutionary timescale. A dream is not unrealistic merely because it is not achieved within one's own lifetime or even that of one's grandchildren. Enthusiasm for one ambitious space dream after another has died out when its supporters came up against the fact that they wouldn't live to see it fulfilled—a reaction that strikes me as all too close to "sour grapes." As has often been pointed out, settling space is not as simple as picking up stakes and moving one's family westward. It requires a very long lead time. During that time, the clock would stop if there were no far-sighted dreamers willing to pay the price of personal disappointment in order to keep it going. The more followers they can attract by offering entertainment, the better; but to suppose that their motivation has no deeper roots is to ignore the essence of what enables our species, or any species, to thrive.

Sooner or later, like an eaglet destined for flight, humankind will break through the invisible shell in which our planet has been confined. It is happening now with the advent of commercial space flight, and the minority with far sight will carry us forward despite reluctance on the part of the majority. In times to come men and women will travel far from this green earth. And then, with our ancestral home at last open to the universe, we will discover our place among the countless worlds of the stars.

Why Does the History of Outlook Toward Space Matter?

The following essay consists of the Preface to the 2022 edition of my 1974 nonfiction book The Planet-Girded Suns *followed by a revised version of the original Foreword to that book. Few of today's space advocates know or care that widespread belief in inhabited exoplanets has a long history dating back to the late seventeenth century. Here I explain what that fact signifies and why I feel that awareness of it is important.*

*

The Planet-Girded Suns is about history, the history of an idea that has been far more prevalent during the past four hundred years than most people realize: the idea that the planets of distant stars are not only habitable, but inhabited. I am not referring to speculation about visitors to our world, either ancient ones or recent UFOs—concepts that arose during the twentieth century which are not accepted by orthodox science. On the contrary, that travel between worlds might ever be possible did not even occur to the scientists, clergymen, and other intellectuals of the late seventeenth through nineteenth centuries who firmly believed that the planets of other suns must have inhabitants. Most argued that God would not have created a world of no use to anyone, but even those who did not put it in religious terms felt that it would be against nature for ours alone to be populated.

Serious opposition to that assumption did not arise until the mid-nineteenth century, and except for a short period in the early twentieth, the belief that there must be many inhabited worlds prevailed. This is not to say, of course, that society at large was aware of the issue or even that stars might be orbited by planets, since most people had too little education to have knowledge of astronomy. But the magazines, newspapers, and poems of the day, directed to the educated minority, made frequent reference to it, as did the writings of such diverse notables as Immanuel Kant, the Puritan minister Cotton Mather, and Benjamin Franklin. It was even included in a few eighteenth-century textbooks for children.

Why should we care today what our forebears believed? Now, what exists on other worlds is a matter for investigation by science. There are countless science books that deal with the issue. This book, originally published in 1974 and updated in 2012 and 2016, contains short sections about the science, but they too are fast falling into the "history" category. Radio astronomers are attempting to detect signals from extraterrestrial civilizations, and more exoplanets are being discovered every day. If you are looking for current scientific information, this is not the place to find it.

What you will find is a perspective on the issue that shows public interest in exoplanets and extraterrestrials is no mere passing fad, a subject assumed (erroneously) to have originated in science fiction. It is a fundamental aspect of human thought. It's significant, I think, that people of past centuries were convinced that other inhabited worlds exist, without any scientific evidence whatsoever. This historical fact reveals that human beings have an instinctive sense of kinship with the wider universe and a desire to see the realms that lie beyond this one small planet—and perhaps, eventually, to go there. Our ancestors conceived of such voyages only in a spiritual sense, as occurring after death. But we who have taken our first small steps into space are aware that our descendants may set foot on the worlds of other suns, and those of us who have faith in such a future believe it to be the destiny toward which humankind has been moving throughout history—a step essential to the long-term survival of humankind. Had this been known to the writers of earlier times who spoke of those worlds with longing, they would wonder at the public apathy toward space travel that prevails just when we stand on the threshold of fulfilling their dream.

And yet, today's apparent apathy may be rooted in something much deeper than is commonly supposed. While updating *The Planet-Girded Suns* for republication, I was struck by a parallel I had not seen before between our time and an earlier era, which puts my worries about our lagging progress in space into a different light. And so in the 2012 of the book I added an Afterword, which has since appeared in *The Space Review*. It argues that when a new concept of the universe arises, it provokes unconscious anxiety in the public that takes time to fade. Just as in the seventeenth century people were upset by the knowledge that the stars are suns scattered in space rather than lights fixed to a nearby sphere, the growing awareness that Earth is not safely isolated from whatever lies beyond makes many of our contemporaries uneasy. Thus the predominant feelings about spaceships are ambivalent. Nevertheless, if a deep, instinctive desire to see other worlds and a conviction that we're not alone in the universe is indeed common among people of all eras, as our history suggests, we can be sure that those who follow us will not turn back from becoming spacefarers.

*

Surprising most people of today is the fact that belief in inhabited extrasolar worlds is not new. The idea was not, as is commonly believed, invented by science fiction writers. On the contrary, it was accepted by the majority of educated people from the late seventeenth century until the early twentieth century. Scientists, philosophers, clergymen and poets wrote a great deal about it. When in the 1850s the head of a well-known college wrote a book suggesting that there might *not* be other inhabited worlds, he published it anonymously because he felt it might damage his reputation—and indeed, most of the book's many reviews were disapproving. A prominent university's magazine declared that plurality of worlds was a subject on which "until now it was supposed that there was scarcely room for a second opinion."

This fact does not appear in history books. Until recently the information was to be found mainly in the books and magazines of past centuries. Famous authors of those eras sometimes mentioned their belief in other worlds, but they spoke of it briefly and casually, thinking it too commonplace an idea to merit much discussion. Most of the writers who went into detail about it are no longer famous. Their books, many of which were bestsellers in their time, have been nearly forgotten. At the time this book was written they existed only in the collections of large libraries, rarely called for, in some cases with bindings so old and brittle that they fell apart in one's hands when one first opened them to read—though by now some have been scanned and are accessible on the Internet.

Such books are not science fiction. Though a few imaginative stories about voyages through space were written as early as the seventeenth century, they were presented as dreams, satire on human society, or fantasy; they did not suggest that space travel would ever become a reality. Not until the late nineteenth century was there any fiction set in the future. The widespread literal belief in extrasolar worlds, on the other hand, was discussed in nonfiction—"popular science" works and also religious ones, reflecting their authors' conviction that God would not have created the stars merely for people on this one small planet to look at. All contain speculation about the inhabitants of other planets that was intended to be taken seriously. Readers did not laugh at speculation of that kind, for none of it—even the portions concerning life on the moon—was contrary to the science of its time. Later scientists, who knew more, looked upon it with scorn. Several generations, the generations that came of age during the years between the twentieth century's two world wars, got the impression that science had always laughed at talk of "space people" and that it always would; not until the 1960s did respected authorities begun to speculate again.

The speculations in old books, and in most modern scientific ones, have nothing to do with UFOs. The question of whether there are inhabited worlds elsewhere in the universe is separate from the question of whether or not any of those worlds' inhabitants have ever visited our world. Nonfiction of past centuries about extrasolar planets does not mention such a possibility. The idea did not occur to anyone until about the time of World War II. Since then, many people—some of whom are scientists—have investigated records of strange objects seen in the past, and have suggested that these might have involved alien visitors. But science considers the existence of other civilizations far more probable than the notion of their representatives' having come here. And during the former period when almost all educated people were utterly convinced that superior civilizations exist, actual contact between the ones of different solar systems was not even imagined.

Until recently searching for the old writings about extrasolar worlds was a little like a treasure hunt: one could not predict just where they would be found, and one had to look in many places without finding anything. Libraries had reference tools that helped, but these tools were only a beginning; often they provided merely clues leading on to other clues. Occasionally they led to a dead end, such as a work of which the only existing copy was in an inaccessible museum. Yet an astonishing number of relevant volumes were available, even before scholars had published the accounts of past writings that now exist. One could go to a library shelf, take down a magazine well over a hundred years old, and turn to an article that thousands of people must have read when it was new—and that never, perhaps, had been looked on by anyone now alive. The wording of such articles may seem quaint, and their authors may have been ignorant of facts that are now known, but the idea expressed is often closer to what scientists are saying today than to what they said when one's grandparents were young.

There are many current science books about extraterrestrial life. This, however, is not a science book. It is the history of an idea. Not all men and women with important ideas are scientists, for science studies only that which can be systematically observed. Long before the invention of the telescope made it possible to observe distant parts of the universe—long before the belief in other worlds became popular—there were men who thought about what might lie beyond Earth. Some had followers, but others were ridiculed or persecuted and at least one was put to death for his theories. Since that time more facts about the universe have been learned; present views of far-off solar systems have scientific foundation. Still, the question of what inhabitants of those solar systems are actually like cannot yet be studied scientifically. When scientists give opinions on it, they are speaking not as authorities but simply as members of the human race, just as their predecessors did. They are expressing not proven truths, but thoughts.

Thoughts about the unknown concern not only science, but religion. For many centuries all speculation about astronomy was inseparable from religion, since the mysteries of the heavens could be explained only in religious terms. Today, when more scientific data can be obtained, there seems to be a firm line between the two. In the past, however, people who drew a line between religion and other affairs placed the subject of other worlds on the "religious" side of that line. Before the twentieth century, few if any separated their personal religious beliefs from their thoughts about what the universe is like. Even those who paid little attention to religion in everyday life considered cosmology—the nature of the cosmos—too unknowable to be viewed as a purely scientific matter.

That astronomical discoveries came into conflict with the religious view current at the time of Copernicus and Galileo is a familiar fact of history. It is often said that learning that the earth moves around the sun lessened people's feeling of central importance. Historians, however, point out that the relation between the earth and the sun was not the real issue. More upsetting was the realization that there are other suns, and therefore, perhaps, other earths—innumerable earths, all of equal importance in the universe. Yet though this was a blow to human pride, before long the public began to look upon the existence of countless worlds as proof of God's power and glory. When no scientific evidence is available, faith of some kind is the only basis for believing in the unseen.

Near the end of the nineteenth century another crisis occurred, one that has not been discussed often. People had been saying for two hundred years that a world would not be created for no purpose, and the only purpose anyone could think of was habitation. Travel from one world to another was not thought possible. So when scientists concluded that the moon and nearby planets are not inhabited, it was natural to start wondering whether the universe is really purposeful. The most common argument for extrasolar life seemed less convincing than before. Furthermore, around the turn of the twentieth century a new theory was adopted about the origin of planets. Astronomers began to think that solar systems came into existence accidentally. Such accidents were considered rare; even among people who still viewed cosmology in a religious way, there were many who abandoned their faith in worlds of other suns.

Today, the opposite situation prevails. Since the mid-twentieth century scientists have believed it is highly unlikely that ours is the only inhabited planet in the cosmos, for solar systems have been considered common—a theory recently confirmed by the discovery of many planets orbiting other stars. The likelihood of sentient species elsewhere is accepted by men and women of differing faiths, and also by those with no religious faith. It is frequently assumed by the latter that discovery of extraterrestrial life would be upsetting to religion. This is not true; there has been little if any conflict since the early seventeenth century and most if not all the religious thinkers who have considered the issue believe that existence of other inhabited worlds is compatible with their faith. (Interestingly, a poll has shown that many people think the idea would be disturbing to adherents of other religions, though not to those of their own.) Yet the former Soviet Union's philosophy of dialectical materialism supports the same idea, as did Russian Cosmists before the Soviet era. In 1958 a Soviet astronomer wrote, "The thesis of the existence of life outside the earth is shared in our epoch in equal measure both by the materialists and by the idealists." There are few issues of such importance on which people with conflicting philosophies can so readily agree.

In the early seventies when I wrote *The Planet-Girded Suns* I was not aware of the Russian Cosmists, about whom little information was then available in English, nor could I locate any information about the views toward extraterrestrial life of religions other than Christianity and Judaism. Since then I have come a across a book dealing with Muslim views based on the Koran. The conviction that our planet is but one of many supporting life appears to be innate in significant numbers of people, regardless of time and place. If this is true, our thoughts about our relation to the universe are not mere idle curiosity, but are basic to our conception of human destiny.

Confronting the Universe in the Twenty-First Century

The following essay, in which I first discussed the parallel between past centuries' response to new views of space and our own, is a slightly-revised version of the Afterword to the 2012 edition of my nonfiction book The Planet-Girded Suns. *It was also published online in* The Space Review. *An update from the 2016 edition of the book has been added to the end.*

*

Like most space advocates, I have long been discouraged by the slowness of our progress in becoming a spacefaring civilization. I have sought desperately for a way to combat the widespread indifference to that goal.

But is it really indifference? Early in this century I came to believe that the past four decades' waning enthusiasm for space travel on the part of the general public is due to something deeper than that. I think, as I said in a 2006 essay at my website, that it is a matter of unconscious fear. In revising my 1974 book The Planet-Girded Suns for its 2012 republication, I became more than ever convinced that this is true—and yet I now feel that our society's current ambivalence toward space may not be the tragedy that I once thought. In historical context, it can be seen as a natural and predictable stage of human progress.

In the seventeenth century, all but an exceptional few resisted the idea of a universe full of solar systems because it shook the foundations of the safe, stable world they knew. The root of their resistance was fear. It is widely recognized that the men in high positions, both religious and secular, were afraid that upsetting the established conception of the physical world would undermine their authority; but it is not that fear to which I'm referring. More fundamental was people's instinctive, inner suspicion that if the universe was larger and less orderly than had been supposed, there was no knowing what threats it might hold.

But this fear eventually faded. The innate human desire to know more of the stars is powerful, and by the eighteenth century the intelligentsia, having gotten used to the idea of many solar systems, were envisioning them with yearning as an embodiment of God's glory. There was, to be sure, a countercurrent; Pascal's admission, "The eternal silence of these infinite spaces terrifies me," is merely the earliest expression of a human reaction that is widespread, though seldom acknowledged. But in Pascal's time and long after, humans were insulated from the universe by a gulf assumed by most to be unbridgeable. For two and a half centuries the same picture prevailed: stars are suns; some or all of them are surrounded by planets, at least a few of which are similar to, or better than, ours; their inhabitants may be the equivalent of angels; and however much we long to see those worlds, they are—by mortals, in any case—inaccessible.

And then, around the middle of the twentieth century, the foundations of humankind's worldview were shaken again. Gradually, by a growing proportion of the population, it was recognized that extrasolar worlds may not be inaccessible after all.

In past eras the distant reaches of the sky had been associated with Heaven; even people who didn't believe in a physical Heaven had absorbed that emotional connotation from the collective unconscious of their culture. The worlds of the stars had been idealized. There had been no reason to imagine devils there—none of the speculators who viewed them as the homes of the angels, or races as superior to humans as angels, had considered the possibility of devils. In any case, the speculation was all very abstract. It had little to do with real life.

Early in the twentieth century science fiction writers, and a few far-sighted writers of nonfiction, began to envision travel between the stars; but the public perceived that as fantasy. The vast majority, if exposed to it at all, did not incorporate it into their conception either of the real universe or of the future of their descendants. The shock of the technologies produced by World War II, however—the V-2 rockets and the release of atomic energy—weakened the distinction between imagination and reality. It is not a coincidence that the first UFO sightings were reported less than two years after the war ended, or that they were followed by the alien invasion movies of the 1950s, which were not, as is commonly said, veiled expressions of Americans' fear of attack by the Soviet Union, but rather the stirring of an unconscious recognition that the universe is very much vaster, and more scary, than most people like to think.

That recognition was effectively suppressed by Hollywood's portrayal of evil aliens in forms that could be written off as nonsense not meant to be taken seriously. But with the launch of Sputnik in 1957, and ultimately with Apollo 8's voyage to the moon in 1968, the public was suddenly jolted into awareness that our access to space—and correspondingly, interstellar travelers' access to Earth—is real.

Much is said about the positive effect of the photos of Earth obtained by Apollo 8, which for the first time showed our planet as a globe, a fragile refuge amid barren surroundings, and thereby launched the environmental movement. The concomitant negative impact—the spread of gut-level knowledge that space is an actual place containing little that's familiar to us and perhaps much that we'd rather not meet—is not spoken of. But it may be no less significant. Could this be one of the reasons why interest in space died so soon after the first moon landing, resulting in the cancellation of the last few planned Apollo missions?

It is undoubtedly the cause of the rise of belief in UFO contacts; and the experience of being abducted by aliens—which in most cases is neither faked nor a manifestation of mental illness, but a perception that emerges in a form indistinguishable from memory—is, in my opinion, an unconsciously-formed metaphor for the unknown terrors that may await us in space. In the past few decades such submerged worries have spread to a wider and more literal-minded segment of society, even to some who consider themselves space enthusiasts. The decline of positive space imagery in science fiction movies and corresponding rise of fantasy and disaster films reveals that space is less appealing to the public, and its nameless potential evils are more frightening, when the universe is open to humans than it was when it could be viewed as an unreachable realm.

This explains much that has been puzzling to space advocates, who have long sought an answer to what happened to the vision that offered such promise and evaporated so suddenly. Expansion into the new ecological niche of space is clearly a new stage of human evolution, yet after brief acknowledgement at the time of the first moon landings, our society as a whole has been blind to this . . . or perhaps not. Perhaps underneath people know it all too well. Even space supporters often feel no urgency about bringing offworld settlements into existence; many dream of them as symbols of a hopeful future, but some dreamers, like almost everyone else, may be reluctant to take the plunge. Relatively few individuals really enjoy the thought of being on the cutting edge of a major step in human evolution, for who knows where that may lead? At the time of Columbus, uneducated people thought adventurous ships would fall off the edge of the world, a prospect they viewed with great dismay; others (according to legend), knowing the world extended beyond their maps, marked the edges with the warning "Here Be Dragons." Figuratively speaking, most people of our time may feel the same way about space exploration.

As members of the baffled minority have bitterly noted, we have spent forty years aimlessly circling Earth in low orbit, with no commitment to venture any further. The false dawn that science fiction author Robert Heinlein postulated as a mere fictional device, not intending it as prophecy, has ironically proven to be a truer portrayal of the post-Apollo era than the vision of space travel by which he and his readers were inspired. By now some people are even claiming (or perhaps hoping?) that the moon landing we once achieved was a hoax. No one who watched that landing foresaw such an outcome; of all the predictions that might have made about the near future, the cessation of exploration would have seemed the least likely. And yet, seen with hindsight, it is not strange.

The assimilation of awareness that this planet is not isolated from the universe is a far more profound upheaval in human thought than the one that aroused so much opposition in the seventeenth century. People of that era were uncomfortable with the idea that the earth moves around the sun, and the sun is only one of many; but they did not suppose that this might have an impact on their personal lives or those of their grandchildren. The idea of contact with other worlds is another matter entirely. Whether we go there or their inhabitants come here, a major shakeup of Earth's civilization is inevitable. Writers who have contemplated this issue have been warning about "culture shock" for a long time, and more recently some have begun to suggest that the notion of hostile aliens may not be as silly as was once assumed. Either way—and even, or perhaps especially, if explorers were to encounter some incomprehensible presence in space that does not fit our present conception of "aliens"—the ways of thinking based on millennia of human tradition would be shattered. Without that foundation the average person's life would be in shambles.

These are not new ideas; they have been discussed for years, both by serious science fiction and by nonfiction, in speculation about the consequences of actual alien contact. But the average person is not a reader of either, and those who do read them—members of the minority attracted to the challenge of space—do not realize how much emotional impact such underlying implications have already had on the population in general. Most people do not want to contemplate the significance of an open universe. They do not let uneasiness about it into their minds; but underneath, as the collective unconscious of humankind absorbs the knowledge, they grasp it, and react with dismay disguised as apathy. It does not occur to them that they might be disturbed by the prospect of space exploration. Rather, they believe that although in theory they want humankind to reach new worlds, it's of low priority compared to the problems of here and now.

Thus the excitement that prevailed at the time of the early Apollo flights was overwhelmed by the emotion aroused by the sight of Earth's beauty and fragility as seen from space, and the focus of the majority turned inward. "Earth must be protected," became the watchword. Protected against what? The threat of nuclear war had been seen with alarm for the past quarter of a century; the competition with the Soviet Union to reach the moon was, after all, a response to that threat. The threat to the environment, though newer, was recognized by some and would eventually have become apparent without the impetus provided by the moon flights. But the sight of Earth as a globe aroused more than the realization that Earth's environment is fragile. It created awareness that Earth itself is vulnerable—vulnerable to whatever strange threats may exist beyond our present knowing.

Beyond what we want to know. And the further humankind ventures into space, the more likely we are to find out.

So the public, without conscious recognition that anything more than economics underlies its reluctance to fund more space expeditions, has succumbed to what perplexed space advocates have interpreted as indifference. It is, I believe, the opposite— far from lacking emotion about other worlds, most people who are in no hurry to see a human presence established there are suppressing ambivalent emotions of extraordinary power. Nothing less can explain the vehemence with which some reject any suggestion that extraterrestrial resources would be of inestimable value in preventing the depletion and pollution of our home planet. Nothing less can account for the fact that Earth-worship has become a religion—sometimes even in the literal sense—to many, or the claim that it is not merely foolish, but morally wrong to advocate the spread of humankind beyond the one small world that overpopulation would otherwise eventually ruin.

What people feel most strongly about does usually find religious expression, and because contrary to a common misconception, traditional religions do not oppose belief in extrasolar worlds, the notion that Earth is the only place of value in the entire universe has given rise to a new one. To be sure, there is an alternate form of moral opposition to interest in new worlds. Ever since space travel became feasible some have opposed it on the grounds that with it, we would be bound to destroy native races elsewhere. In popular culture this, the reverse of the alien invasion idea, is reflected in the hit movie Avatar, in which humans are portrayed as evil invaders of an alien paradise. The fallacy in the film is not the theme that taking over an inhabited world would be wrong; of course it would be—no modern space advocate would countenance establishment of enclaves on inhabited planets. (Attitudes have changed radically since Heinlein's fiction did so.) Only rationalization of the fear of unknown realms causes people to emotionally accept the premise that this would happen.

I sometimes wonder if the widespread conviction that the public no longer cares about space may also be a rationalization.

After all, if the public has lost interest, then one need not question one's own reluctance to support space programs—even if underneath one feels a conflicting impulse to rejoice in the awesome spectacle of a launch. A politician need not feel any obligation to fund them, no matter how many activists explain why they should be funded. When a candidate for the 2012 presidential nomination proposed returning to the moon, the other candidates did not argue against the idea; they simply ridiculed it. Ridicule is a convenient weapon against buried feelings that one does not dare to acknowledge.

Over and over during the past four decades, space advocacy organizations have flared up, shown promise, and then died; only a few have lasted and their membership has never approached a fraction of the people belonging to environmental organizations. Time and time again space projects have been initiated, only to be cancelled before they came to fruition; even robotic Mars exploration, about which the public did show enthusiasm, has now been cut back. Nominal space supporters may well be influenced by the same inner apprehension as the rest of the population. There are, of course, many dedicated activists who have done, and are still doing, everything in their power to get us off the ground; but support for them has been ephemeral.

We are driven by the urge that impelled generations of our forebears to dream of other worlds, and yet whenever we are on the verge of a step toward them, we pull back. My own expectation of seeing a human base on Mars within my lifetime has waned from certainty in the Apollo era to mere hope, to doubt, to despair, and finally to resignation. Yet as I have said above, I do not now view this as tragic. A tragedy is a turn of events that should have been avoidable, and I no longer believe that the present hiatus in space travel could have been avoided. When in updating The Planet-Girded Suns I reread it after passage of time, I was struck by the parallel between the cusp of the twenty-first century and that of the seventeenth; and I see that our slow progress toward becoming a spacefaring species is no less a stage of human evolution than any other overpowering change of perspective. Stages of evolution cannot be skipped or hurried; that, after all, is a major theme of my novels.

To be sure, the stage dealt with in one of those novels—the Critical Stage described in The Far Side of Evil—is characterized by great danger. When I conceived the story (in 1956) and wrote it (in 1970), I seriously believed its premise that the point in history at which a planetary civilization develops space travel capability coincides with the point at which it becomes capable of self-destruction, and that the two are mutually exclusive alternatives. By the time I revised it for republication in 2003, our lack of commitment to space had forced me to realize that the Critical Stage lasts longer than the few years I originally assumed, and to the danger of nuclear war I added mention of such threats as pollution, depletion of resources, and widespread terrorism. I still believe that humankind will remain in peril unless and until we make the effort to expand beyond this world. But I do not think that doing so is merely a matter of talking the public into it.

While this is a discouraging realization, in another sense it is cause for hope. It shows that nothing has yet gone wrong, that our retreat from the outward thrust begun by the first flights to the moon is not a sign of human failing, but a normal and inevitable reaction to those flights. Just as it would have been irrational to expect most contemporaries of Bruno or Galileo to embrace a new conception of humankind's place in the universe, it is unreasonable to expect it of ours. It takes time for a culture's fear of the unknown to fade.

Yet it does fade, and the human longing to reach the stars is stronger and more lasting than the hesitancy brought on by frightening new worldviews. People were drawn to ideas about stars long before the concept of solar systems arose. We have gazed at them since the dawn of history, and nothing short of a fundamental change in human nature could make us turn away. This is not the place to discuss my conviction that the collective unconscious of humankind contains perceptions of reality not acquired through experience, and that the sky has—in all cultures—been a symbol of human aspiration precisely because our future survival depends on going there. Understanding of our present situation does not depend on belief in destiny. It is enough to compare the predominant feelings toward space of the eighteenth century with those of the early seventeenth, and those of the 1960s with those of the 1930s, to see that a sense of relationship to the stars always returns.

To some space advocates, especially those young enough to hope for personal involvement in establishment of an offworld settlement, what I've said here may seem unduly pessimistic. It's natural to shrink from awareness of what one cannot have. Many who once believed in orbiting space colonies abandoned the idea from frustration when it became apparent that it wasn't in the cards for such colonies to be built in the immediate future, just as in the late nineteenth century people stopped envisioning extrasolar worlds when they stopped believing that they would see them on the way to Heaven when they died. No doubt frustration, as well as fear, has played a part in the public's loss of interest in space.

Possibly I see now what I did not see earlier only because I am past the age of hoping that I will live to watch humans travel beyond the moon (or even that far, unless a privately-funded expedition gets there). Yet to me it is reassuring to realize that, barring some major disaster to Earth's civilization, the present unwillingness to accept the challenge of space will pass. That it is not a aberrant prolongation of the Critical Stage as I have long feared, but simply a pause to prepare humankind for the momentous journey ahead.

POSTSCRIPT, JULY 2016

Reading this essay four years after its first publication, I feel more than ever that what I have said explains the reluctance of the public to support a major effect toward human expansion into space. The danger in confining our species to a single planet despite the possibility of worldwide disaster is too obvious for me to doubt that there are deep-seated psychological reasons for the public's refusal take it seriously. Perhaps, in addition to the factors I have already mentioned, people do not want to think to think this danger is real, and are therefore unwilling to acknowledge it by accepting the need for an offworld presence. And yet a great deal is said about threats to Earth, and the majority the people most concerned about them are the ones least apt to push for space colonization. So underneath, suppressed fear of the vast unknown universe must be stronger than worry about asteroid strikes, nuclear war, excessive climate change, or runaway biotechnology.

But now, there is greater cause for optimism about progress in space than there was four years ago. Now, privately-owned ships are delivering supplies to the space station, manned flights are soon to follow, and a privately-funded mission to Mars is in the planning stages. This is the first truly promising development since the Apollo era. I have always known that private investment alone could bring about a real breakthrough, but I feared that investors would not grasp the enormous economic potential of space activity and extraterrestrial resources until far in the future. I'm thankful that it has proved to be a mistaken fear.

And so I feel sure that we are on track, and that the gradual extension of our species' range will occur just as it always has, through the efforts of individuals who take risks for the sake of what they personally believe, or hope to gain, and thereby bring about gain for all humankind. That is how it has worked in the past—with the earliest tribes that ventured beyond their territory, with the explorers who sailed from Europe to the New World, and with the later pioneers who migrated to the American West. It is not necessary for society as a whole to support such efforts. Had widespread enthusiasm been needed for Columbus to sail, he would never have left Spain. It is only looking back that we see the full value of what individuals strive to do, and this will be as true of progress in space as it has been on our home world.

A colony on Mars may not be established in my lifetime, but I am confident that it will be during the lives of most people who are now living. Whether they like the idea or not does not matter, any more than public opinion mattered at the time Bruno first suggested that there are planets circling other suns. Ultimately we will reach some of those exoplanets, and long before we do, spacefaring will seem as natural and desirable as air travel does today. In the light of human history, we need not be discouraged by the fact that the time is not yet ripe for this outlook to prevail.

Space and Human Survival

The essay below is a statement that I originally wrote in 1994 for the students in an online Media Studies course I taught for Connected Education about Space Age Mythology. (I've since made a few minor modifications.) It wasn't part of the course material, but simply explained personal views I had often referred to in online discussions. It has been posted at my website since 2001 and has been read by many space advocates who aren't familiar with my novels.

*

Until space travel became a reality, the reason most commonly offered for believing our survival depends on it was that our species will need to move elsewhere in order to survive the ultimate death of our sun, or the possibility of our sun turning into a nova. Scientists now believe that these specific scenarios won't happen; but the sun will eventually become a red giant, which as far as Earth is concerned, is an equally disastrous one. This is not of such remote concern as it may seem, as I will explain below. However, it surely is a remote event, billions of years in the future, and I don't blame anyone for not giving it very high priority.

A more urgent cause for concern is the need not to "put all our eggs in one basket," in case the worst happens and we blow up our own planet, or make it uninhabitable by means of nuclear disaster or perhaps biological warfare. We would all like to believe this won't happen, yet it is hardly an irrational fear. Peace with Russia may have drawn attention from it, yet there are other potential troublemakers, even terrorists; the nuclear peril is not mere history. Furthermore, there is the small but all-too-real possibility that Earth might be struck by an asteroid. We all hope and believe our homes won't burn down, and yet we buy fire insurance. Does not our species as a whole need an insurance policy?

Even Carl Sagan, a long-time opponent of using manned spacecraft where robots can serve, came out in support of space colonization near the end of his life, for this reason; see his book

Pale Blue Dot. And in an interview with Britain's newspaper Daily Telegraph, eminent cosmologist Stephen Hawking said, "I don't think that the human race will survive the next thousand years unless we spread into space. There are too many accidents that can befall life on a single planet." Hawking is more worried about the possibility of our creating a virus that destroys us than about nuclear disaster. However, he said, "I'm an optimist. We will reach out to the stars."

My novel The Far Side of Evil is based on the concept of a "Critical Stage" during which a species has the technology to expand into space, but hasn't yet implemented it, and in which that same level of technology enables it to wipe itself out. The premise of the book is that each world will do one or the other, but not both. I have believed this since the early fifties, when there was real danger of nuclear war but no sign of space travel. When the Russians launched Sputnik in 1957, my reaction was overwhelming joy and relief, because I thought that at last our energies were going to be turned toward space exploration. I felt that way through the era of Apollo. Since Apollo, as public support of the space program has waned, my fears have grown again, because I don't believe that a world turned in on itself can remain peaceful. A progressive species like ours has a built-in drive to move forward, and that energy has to go somewhere. Historically, when it was not going into mere survival or into the exploration and settlement of new lands—which is the adaptive reason for such a drive—it has gone into war.

This is the price we pay for our innate progressiveness. I know that it is now fashionable to deride the concept of progress, and certainly we cannot say that progress is inevitable. It surely doesn't characterize all change in all areas of human endeavor. Nevertheless, overall, the human race as a whole advances; if it did not we would still be cavemen. This is what distinguishes our species from all others. And like it or not, this drive is inseparable from the drive toward growth and expansion. Many successful species colonize new ecological niches; this is one of the fundamental features of evolution. When a species can't find a new niche, and the resources of the old one are no longer sufficient, it dies out. If the resources do remain sufficient, it lives, but is unchanging from era to era. There are no cases in biology of progressive evolution unaccompanied by expansion.

Thus it will ultimately be essential for us to colonize space. People sometimes object to the term "space colonies" on political grounds and for this reason NASA, along with some others, prefers the term "space settlements." The objection, however, strikes me as invalid. To be sure, "colonization" does have some bad associations, since on Earth it always involved taking over the land and/or culture of indigenous inhabitants—but that is precisely what a space colony would not do! Nobody, to the best of my knowledge, advocates colonizing inhabited planets, even if we should ever find any. The idea of expanding into space is to abandon our dependence on zero-sum games. A more accurate precedent for the term "colonize" in the space context is its meaning in biology: the establishment of a species' presence in a new ecological niche.

Earth's eventual shortage of resources is an even more crucial reason for expansion into space than the danger of its destruction. It's obvious that this planet cannot support an expanding population forever. Most people who recognize this fact advocate population control to the extent of "zero population growth." I do not; I believe it would be fatal not only for the reason explained above, but because if it could be achieved it would result in stagnation. I do not want a world in which there can be no growth; growth leads to intellectual and artistic progress as well as to material survival. Furthermore, I do not believe it could be achieved. The built-in desire for personal descendants is too strong; that is why our species has survived this long, why it has spread throughout the entire world. Moreover, the biological response to threatened survival is to speed up reproduction, as we can see by the number of starving children in the world. If we tried to suppress population growth completely, we would have either immediate violent upheaval or a period of dictatorship followed by bloody revolution. Ultimately, we would reduce the population all right; we would decimate it. That may be "survival" but it's surely not the future we want.

We do not want even the present restriction on resources.

Currently, some nations live well while others are deprived, and it's asserted that even those with the best access to resources should stop using them up—the underdeveloped nations, under this philosophy, are not given the hope of a standard of living commensurate with the level our species has achieved. Will the Third World tolerate such a situation forever? I surely wouldn't blame them for not wanting to. And neither do I want the rest of the world reduced to a lower level of technology. Even if I had no other objection to such a trend, the plain fact is that a low level of technology cannot support the same size population as a high level; so if you want to cut back on technology, you have to either kill people outright or let them starve. And you certainly can't do anything toward extending the length of the human lifespan. This is the inevitable result of planning based on a single-planet environment.

If there is pessimism in Earthbound science fiction (which its most outstanding characteristic), these truths are the source of it. I have not seen any that denies any of them; pop-culture SF films reveal that what people grasp mythopoeically about such a future involves catastrophic war, cut-throat human relationships in overcrowded cities, and a general trend toward dehumanization. To give only a few examples, Soylent Green postulates cannibalism and Logan's Run is based on the premise that everybody is required to die at the age of thirty. The destruction of the world's ecology is a basic assumption of such fiction— which is natural, since in a contest between a stable biosphere and personal survival, humans will either prevail or they will die.

Myths showing these things are indeed part of the response to a new perception of our environment: the perception that as far as Earth is concerned, it is limited. (A basic premise of the course during which this was written was that all myth is a response of a culture to the environment in which it perceives itself to exist.) But at the rational level, people do not want to face them. They tell themselves that if we do our best to conserve resources and give up a lot of the modern conveniences that enable us to spend time expanding our minds, we can avoid such a fate—as indeed we can, for a while. But not forever. And most significantly, not for long enough to establish space settlements if we don't start soon enough. A significant human foothold in space is not something that can be gained overnight.

I have called this stage in our evolution the "Critical Stage"; Paul Levinson (who was the Director of Connected Education, the organization for which I taught the course during which this was written) uses different terminology for the same concept. He says that we have only a narrow window to get into space, a relatively short time during which we have the capability, but have not yet run out of the resources to do it. I agree with him completely about this. Expansion into space demands high technology and full utilization—although not destructive utilization—of our world's material resources. It also demands financial resources that we will not have if we deplete the material resources of Earth. And it demands human resources, which we will lose if we are reduced to global war or widespread starvation. Finally, it demands spiritual resources, which we are not likely to retain under the sort of dictatorship that would be necessary to maintain an allegedly sustainable global civilization.

Because the window is narrow, we not only have to worry about immediate perils. The ultimate, unavoidable danger for our planet, the transformation of our sun, is distant—but if we don't expand into space now, we can never do it. Even if I'm wrong and we survive stagnation, it will be too late to escape from this solar system, much less to explore for the sake of exploring.

I realize that what I've been saying here doesn't sound like my usual optimism. But the reason it doesn't, I think, is that most people don't understand what's meant by "space humanization" (a term current at the time this was written that has since fallen out of use). Some readers are probably thinking that space travel isn't going to be a big help with these problems, as indeed, the form of it shown in today's mythology would not. Almost certainly, they are thinking that it won't solve the other problems of Earth, and I fear all too many may feel that the other problems should be solved first.

One big reason why they should not is the "narrow window" concept. The other is that they could not. I have explained why I believe the problem of war can't be solved without expansion. The problem of hunger is, or ultimately will be, the direct result of our planet's limited resources; though it could be solved for the near term by political reforms, we are not likely to see such reforms while nations are playing a zero-sum game with what resources Earth still has. Widespread poverty, when not politically based, is caused by insufficient access to high technology and by the fact that there aren't enough resources to go around (if you doubt this, compare the amount of poverty here with the amount in the Third World, and the amount on the Western frontier with the amount in our modern cities). Non-contagious disease, such as cancer, is at least partially the result of stress; and while expansion won't eliminate stress, overcrowding certainly increases it. The problem of atmospheric pollution is the result of trying to contain the industry necessary to maintain our technology within the biosphere instead of moving it into orbit where it belongs.

What about the growing problem of international terrorism? Unfortunately, it is exactly what can be expected in a Critical Stage civilization: one that has outgrown its home world but has not yet directed its energies into moving beyond, and in which the evil actions of a few individuals can affect the entire planet. In one way this is a hopeful view; it reflects my belief that the threats we face are not signs of something having gone wrong with our species' evolution, but natural ones with which we must deal. But time is running out. To let the fight against terrorism distract us from developing space technology would, in my opinion, be self-defeating, just as would attempting to solve other worldwide problems while remaining confined to Earth.

In short, all the worldwide problems we want to solve, and feel we should have solved, are related to the fact that we've outgrown the ecological niche we presently occupy. I view them not as pathologies, but as natural indicators of our evolutionary stage. I would like to believe that they'll prove spurs to establishing a human presence in space. If they don't, we'll be one of evolution's failures.

But by establishing a presence, I don't necessarily mean exploring and settling other worlds

*

When we think of space exploration, we usually think of its goal as "To seek out new life and new civilizations, to go where no [hu]man has gone before." That's what excites us and inspires awe, in some of us at least, and that's certainly the fountainhead of our mythology. Personally, I believe that from the evolutionary standpoint the joy of exploration is a built-in factor for preservation of the species, just as is the joy of sexual love. But, as our feelings about sexual love mean much more to us than biology and have been the source of many great achievements of our civilization, our exploratory instinct means more than survival. The discovery of new lands has always led to a renaissance in the arts and in intellectual progress, and the same will be true of expansion into space. This process is an aspect of our creativity. We do not explore because we want to survive, any more than we make love because we want to survive; survival is only a byproduct.

However, at this stage of our evolution we have run into a problem with the process. Columbus explored because of his personal urge to do so, and both the Renaissance and human survival followed. (Explorers of some sort were essential to survival—imagine what would have happened if our species had been forever confined to the single site where it diverged from its hominid ancestors.) It was difficult for explorers to get money for ships, but each had to talk only one backer into it; Columbus, according to legend, convinced Queen Isabella. Settlers could move into new lands with their personal resources alone, as Americans did when they loaded their belongings into wagons and set out on the Oregon Trail. Both explorers and settlers were laughed at by people who didn't share their views; it didn't matter. They went anyway. It wasn't necessary for their culture as a whole to decide that it wasn't a waste of money

Not so with space humanization. We can't rely on the drive toward exploration because, by the population at large, it's not considered a top priority. It never was, in any society. If the people of Columbus' time had had to vote to tax themselves in order to fund his ships, he wouldn't have gotten anywhere; most of them felt he would fall off the edge of the world, and even the educated minority, who knew better, felt there was better use for their money. Even in that era, the most altruistic would no doubt have preferred to give Isabella's jewels to the poor. There were some myths, travelers' tales, about riches to be found in new lands; but just as in our time, rational, hardheaded skepticism ruled the majority.

Yet purpose as expressed in mythology is the opposite of rationally-derived purpose. Mythology reflects what we feel, not what we know consciously. Thus Space Age mythology shows us why we would like to explore space, but not why the majority should be willing to pay for it. It shows our dreams, but not what science reveals as the concrete advantages. People who enjoy the mythology don't need hardheaded justification (though even they are often unwilling to vote on the basis of their feelings), while those who don't enjoy it are apt to judge the whole issue of space humanization on the basis of admittedly-impractical mythic metaphors.

It is true enough that we can't solve the problems of Earth by setting forth in starships like the Enterprise, or by interplanetary travel at all. From an economic standpoint, a trip to Mars is not the best way to begin the process of expansion (though it's certainly a later goal, and I support doing it first on the grounds of its effect on the public imagination).The basic ideas of space humanization are (a) to make use of extraterrestrial resources to supplement those of Earth; (b) to move heavy industry off Earth, where it pollutes and where energy is expensive, into orbit, where energy is cheap; and (c) to provide large areas of living space to which people can eventually move (not to "ship extra people into space," which as critics are quick to point out, would not work, but to make room for new people to be born without increasing Earth's population). Only in this way can we get the resources we need both for preserving Earth's biosphere and for eventually building starships.

If you have not heard of this scenario before, it's likely to strike you as impossible, impractical, or prohibitively expensive, if not all three. It certainly isn't what mythology has thus far prepared us for. And yet, we have had the technological capability to begin this process since the late 1970s and it's not nearly as costly as the exploration of a planet without prior space industrialization. The key to it is that we wouldn't try to lift the components of space habitats up from Earth. We would use raw materials from the moon and asteroids, and build solar power satellites in orbit. The power would then be beamed to Earth, where it would be cheap enough to lift the Third World out of poverty (many people in the Third World spend a large share of their time and/or income on firewood, and in so doing, destroy forests). Products of space-based industries would be shipped down to Earth, not lifted up out of its gravity well. Some scientists feel that enough food could be raised in orbit to ship food down, as well. And meanwhile, the space-dwellers producing all these things cheaply for Earth would be getting rich, because they would not be citizens of Earth nations; they would be citizens of their own orbiting colonies, entitled to the full proceeds of their labor. Eventually, they would be rich enough to fund interstellar expeditions. And their living conditions would not be what you're imagining if you're picturing Deep Space Nine. Orbiting colonies—probably the most difficult concept to understand if you haven't seen any of the artists' renditions—would be little worlds built from extraterrestrial materials, with the living space on the inside. They would be complete biospheres with trees and lakes and gardens, much less crowded and less sterile than New York City. Many of their advocates have said that having once lived that way, humans would never want to live on the surface of a planet again, and that if they traveled to a new planet, they'd go to its surface only to explore.

Much of this, in particular the design of the colonies, is the vision of Gerard O'Neill, formerly professor of physics at Princeton and until his untimely death, president of the Space Studies Institute which he formed to research the engineering details of the scenario. His book The High Frontier is a classic that should be read by everyone serious about space settlements. At one time there was an active citizen's group, the L-5 Society, dedicated to his ideas, but it has merged into the National Space Society. He testified before Congress many times and was recognized as an expert on the future of space, though his specific proposals weren't taken seriously by enough people to count.

NASA did two studies of his orbiting colony concept. But of course, though it was entirely feasible from the technological standpoint, it was not feasible politically or financially, at least not in this country. Japan and India were more enthusiastic and I won't be at all surprised if the first orbiting colony turns out to be Japanese.

Most space experts don't advocate anything as ambitious as O'Neill Colonies. It's not likely that space industrialization will proceed that rapidly. But we could do it in stages. We could and should build the solar power satellites (studies that have "proven" them impractical have been based on the assumption that materials would be lifted from Earth; use of lunar materials would make them cost-effective). And we could certainly start utilizing the too-long-abandoned moon. But the American people seem blind to the need to do so, and while private corporations could ultimately get rich by doing it, it's a very long-term investment.

At the time I wrote this essay, I believed that orbiting colonies of the kind proposed by Gerard O'Neill were the most practical first step in establishing a major human presence in space and that they could be built soon if enough effort and funding were devoted to it. We now know that they are a long way ahead and are not the top priority. That does not invalidate the basic facts about the need to make use of extraterrestrial materials and energy and to move polluting industry into orbit in order to preserve Earth's environment. I still think there will be orbiting colonies someday, but other things, such as a permanent base on the moon and/or Mars, and perhaps the mining of asteroids, will precede them.

Of course, I too am excited by the long-range possibilities of galactic exploration shown in Space Age mythology. Paul Levinson has a lot to say about the infinity of the universe and how, in principle, our species has access to its infinite resources and the infinite extension of intelligence this will make possible. I agree wholeheartedly (except that unlike him, I believe we will meet other intelligent species someday). But none of this can happen unless we survive long enough to make it happen. And we can't survive that long, in my opinion, unless we take the necessary steps to get from here to there. This is why I believe the most crucial function of our new mythology, and the one with the greatest adaptive value, is expression of the idea that people belong in space.

POSTSCRIPT (2019)

While I still strongly believe most of I said in this essay, I now feel that it is neither possible nor necessary to convince the public. The majority is not emotionally ready to face the idea of an environment larger than our home planet, and the advances will be made by the minority with exceptional vision, as they always have been throughout human history. When I stated above that "at this stage of our evolution we have run into a problem with the process," implying that our situation is different from the past when exploration didn't depend on society as a whole deciding that it wasn't a waste of money, I was wrong. Entrepreneurs with vision are already beginning to send ships into space, and what the general public thinks of their ventures doesn't matter. If you are a space advocate who has been discouraged and depressed by the slowness of our progress in space in the half-century since the Apollo 11 moon landing, take heart from the success of commercial space flights.

The Only Sensible Way to Deal with Climate Change

The essay below is an expanded version of a section of my essay "The Future of Being Human" in my 2020 ebook by that name. My view of this issue differs from the prevalent one, and I'm aware that some readers will think that what I propose is at best impractical and at worst a distraction from what in their opinion is a better approach to a situation they see as preventable. But though I've always been opposed to doomsaying, I feel strongly that the public's unrealistic view of climate change will prove disastrous unless constructive steps are taken to prepare for it.

*

Currently there is a great deal of emotional and political turmoil over evidence that the climate of Earth is changing. It has changed before, but that was when no humans were around to be surprised. Change is a law of nature, but to most people "nature" means what they or their recent ancestors have been used to; they seem to think the state familiar to them is, or should be, guaranteed. We are learning that it is not, but this strikes most as so foreign to what they expect of the universe that they assume change must be their fault. Perhaps this is less frightening than realization that, short of science fiction speculations about terraforming, no species has the power to alter the environmental conditions of a whole planet.

The present crusade to stop climate change worries me deeply, not because I'm a "denier"—I certainly don't deny that detrimental changes in the climate are likely—but because the action commonly advocated may well do more harm than good. It makes people feel that the problem can be eliminated when in fact it cannot, and prevents them from considering the real solution to it. Adaptation to a changed climate will demand more of humankind than a futile attempt to turn back the clock that if successful, would make us even less able to cope with climate change than we are now. It will require us to move forward by the means that has made us a successful species—the overcoming of obstacles through development of new technologies.

Many people today think that if we don't lower our standard of living, climate change will ruin our planet's environment. In my opinion this particular fear is greatly exaggerated. Contrary to the widely promoted theory that climate change is due to human activity, some scientists believe it is entirely, or almost entirely, the result of natural processes. The idea that we are responsible for it is popular because it offers false hope that we can stop or reverse it, and because it provides an opportunity to vent feelings about it by placing blame.

That said, it would be wise to reduce pollution of the atmosphere by greenhouse gasses whether it is a significant cause of climate change or not. Furthermore, competition for fossil fuels is a major factor in international conflicts, and the supply of such fuels will eventually run out. The need to phase out our use of them is indisputable, even if it has no impact on the climate. But handling the problem of coming climate change has higher priority.

Whatever the reason, Earth is getting warmer, which may cause a rise in sea level and frequent droughts, among other effects. We can no more prevent this than we can prevent earthquakes. We may someday be able to control the weather in specific locations on a short-term basis, but the only solution to problems with the climate is to develop technologies for adapting to it, as humans have been doing since we first built shelters and learned to use fire. That is how evolution works; when environmental changes occur, a species adapts or it dies out. Through extragenetic evolution, we can choose to adapt before our survival is endangered.

Assuming that the economy is not damaged by premature abandonment of existing power sources, by the time climate change is sufficient to threaten the well-being of people in industrialized nations changes in lifestyle made possible by artificial intelligence (AI) will have begun, and the use of AI will lessen its impact. People are going to be moving away from cities, and climate change will accelerate the process. This does not mean that no one will become homeless, but that situation will have to be dealt with anyway, merely from the upheaval accompanying the introduction of AI. Moreover, AI will lead to major improvements in agriculture and food distribution that will mitigate the effects of drought.

However, the problems that will be caused by global warming will be much greater in Africa and Asia than they will be in America. At worst, a rise in sea level could force Americans to move out of waterfront cities and abandon beach homes, but in Asia large populations live in low-lying coastal areas and would have nowhere to go. Developing nations do not have the technology for dealing with high temperatures, such as air conditioning, that are already common in industrialized ones. Nor do their traditional methods of agriculture and food storage provide much leeway in case of extreme weather.

So one of the best ways of adapting to climate change would be to help the people of developing nations obtain modern technology and AI devices. But how could they afford them and the infrastructure required to make them work? I see only one solution, and it's something we should have begun long ago for the welfare of everyone in the world. We need to develop space-based solar power—as soon as possible, and on a large scale.

The key to raising living standards in developing nations is power. About three billion people, mostly in Africa and southern Asia, have no access to electricity. They must use firewood, which they often spend hours gathering, for cooking and heating, and smoke-induced diseases are responsible for the death of 4.3 million people every year. Use of electronic devices is out of the question for these people, but more than that, they need power merely to gain utilities standard in the rest of the world. Yet industrialized nations rely mainly on oil and coal for generation of power, which is a major source of pollution and which will not last indefinitely. It can't be made available worldwide, nor can renewable sources such as Earth-based solar power meet the need.

Since the 1970s it has been known that solar power could be collected by satellites in space and beamed to Earth—to all parts of Earth. This would not only provide plenty of relatively inexpensive power to the nations that launched the satellites, but could raise the rest of the world out of poverty. It would enable the people of poor countries to adopt a lifestyle appropriate to the twenty-first century. Many experts have studied proposals for such a system, but it has generally been considered too expensive a project to undertake. A few, notably the late Gerard O'Neill, have pointed out that it would not be too expensive if materials obtained from the moon were used for construction of space stations in order to save the cost of lifting them out of Earth's gravity well. However, since the moon has been neglected for the past half-century, we will have to develop space-based solar power (SBSP) without the benefit of lunar resources.

This is the only option that makes sense. And it will provide more than power—it will make much-needed water accessible, too. Drinking water is not easy to get in rural areas of developing nations; in Sub-Saharan Africa 14 million women and over three million children spend more than 30 minutes a day carrying water for their households, often from ponds or streams. Water may become scarce in America, too, if climate change adds to the need for irrigation of crops. Already some cities, such as Los Angeles, depend on piping water from distant rivers; if those rivers dried up or an earthquake cracked the pipes, millions of people would die of thirst. And population growth in itself is threatening the water supply. Sooner or later it will be necessary to desalinate seawater. So far, that has been too costly; but with cheap space-based solar power it could be done.

Power satellites will also make possible the utilization of extraterrestrial resources by private industry. Entrepreneurs like those now building spaceships will mine materials from the moon and asteroids and use the profits to expand their operations in space. Eventually, through the use of AI, manufacturing will be done in space as well, and industry that pollutes will be moved out of the atmosphere into orbit where it belongs. This process could restore Earth to its natural beauty, though that won't occur until far in the future.

Space advocates have been arguing for SBSP for many years, and very recently (as of 2022) considerable attention has been paid to its advantages by governments, especially China, Japan, the US and the UK. Several proposals have moved past the planning stage; China now expects to have a megawatt-scale solar power test station operational by 2030. But these are extremely expensive experimental projects. No one, least of all the United States, has made a commitment to providing space-based solar power on a large scale. We have had the technological ability to do it for decades, but it would cost so much money that it's taken for granted that such a commitment would not be feasible. It should be remembered that it was once thought that going to the moon wouldn't be feasible. In any case, the cost must be viewed in relation to the tremendous cost to society of coping with a change in Earth's climate without a sustainable source of power.

The trouble, of course, is that even most space advocates don't think of SBSP as a solution to climate change, and the general public is emotionally focused on the futile and perhaps-damaging effort to make climate change go away. And this is a very dangerous situation. Even with maximum effort, it will take a long time to develop enough solar power satellites to meet the world's need for energy. If we don't start now, and devote all possible human and financial resources to it, the impact on our civilization will be devastating. We may lose the technology that would make it possible.

The space advocacy community needs to adjust its priorities; building bases on the moon and going to Mars are vitally important, but not as critical as providing enough power to prevent widespread loss of life due to climate change and maintain Earth's technological capabilities. Governments need to set new priorities—to stop squabbling over the political significance of climate-change policies and recognize that if and when climate change becomes a crisis, we must be prepared with an energy source that will enable us to deal with it. Above all, the public needs to be informed about the inevitability of changes in our environment and what it will take to mitigate them. Yes, devoting a large share of the world's economic resources to building power satellites would require sacrifices—but would that be worse than losing those resources through an attempt to survive without a sufficient source of power?

Some years ago, speculating about how advanced extraterrestrials might then view humankind, Nigel Calder wrote: "Fecklessness might be the main theme of [the aliens'] report on . . . beings who have mastered a lot of physics, chemistry and biology and yet let their children starve—while all around their planet the energy of their mother star runs to waste in a desert of space." The sun's output is part of our environment, just as the resources of our planet are, and unlike them it will not be depleted for about five billion years. We have always been dependent on it and in the future we will be dependent on using it more fully. It would indeed be irresponsible for us not to do so.

The large-scale use of energy and/or materials from space will be the first major step in extragenetic evolution since our prehistoric ancestors began using metal. It will mark the transition of humankind from a planet-bound species to a spacefaring one, a transition that is essential to the long-term survival of our species. If climate change proves to be the spur needed to make this happen, then like many other apparent misfortunes in our history, it may someday be seen as a blessing.

Update on the Critical Stage:
The Far Side of Evil's Relevance Today

This 2017 essay explains why I feel my novel The Far Side of Evil *should not be called outdated, as it sometimes has been, and also presents my recent thoughts about the theory of the Critical Stage that underlies that novel. Because it is highly relevant to my view of the importance of space, I'm including it in this book as well as in my collection of essays about my writing.*

*

The central idea of my 1971 novel *The Far Side of Evil* is one that came to me 1956: my theory of the Critical Stage, the time between a planetary civilization's development of the means to destroy itself and a commitment to expand beyond the single planet where such destruction would wipe it out. The same level of technology that makes one possible also permits the other, and in my view they are mutually exclusive alternatives—a world will remain in the Critical Stage until one or the other happens. This supposition is as valid in my eyes as ever, although my conception of the Critical Stage has changed over time.

It's frustrating to me that many readers feel that the novel is outdated, and that it therefore seems irrelevant to today's world. I addressed this in the Afterword to the 2003 revised edition, but some people say the new version, too, is dated, so evidently I failed to revise the original text successfully. That's too bad, as I feel the book is even more relevant today than at the time of its initial publication.

The political situation in the story was never meant to parallel current events on Earth; it is comparable to our world as it was during the early fifties, not the seventies when the first edition appeared. After all, I wrote an initial draft of portions of it a year before the launch of Sputnik, an event that to my surprise and relief made it impossible for the planet portrayed to be our own. The story is not about politics, although its setting—the conflict between dictatorship and freedom—is universal and applies to all eras. As far as the story is concerned, that conflict is merely a plot device; so the fact that we no longer have two superpowers on the verge of nuclear war in no way dates it. As I said in the Afterword, some readers thought I used space fiction as a vehicle for political commentary when in fact it was the other way around: I used political melodrama to dramatize my ideas about the importance of traveling into space.

When the book was written, I assumed a world's Critical Stage is short. (Yes, I believe the theory applies to worlds other than ours, just as some scientists now believe that one goal of seeking interstellar radio contact is to find out how long an average planetary civilization lasts before self-destruction.) At the time of the Apollo moon landings, most people thought that nuclear war was likely to occur in the near future, but that if it didn't, we would continue to make rapid progress in space exploration. Since personally, I had believed since the early fifties that devoting a society's energy to space travel puts an end to the danger of a catastrophic nuclear war, I described the Critical Stage in those terms. And in fact, some evidence was provided by the space race with the Soviets, which absorbed money and effort that would otherwise have been spent on a more destructive competition.

As time passed, however, it became clear that my theory was a gross oversimplification. I tried to update it in the 2003 edition, pointing out that there are more dangers to a planetary civilization than nuclear war and that mere development of space travel capability, without a major commitment to establish settlements on other worlds, is not enough to eliminate them. But it's a novel, not a philosophic treatise, and it was being issued by the publisher's "children's book" department (although it's inappropriate for readers below high school age), so I wasn't able to elaborate enough to clarify the relevance to today's world.

Readers say to me that we have space travel yet are still in danger, and that's true. But we haven't made use of our space travel capability. Expansion into space prevents a civilization's destruction by two means: first, by constructively channeling the energy that would otherwise have gone into war, and second, by establishing footholds that can survive even if a species' home world is devastated—which in principle can happen through a natural event such as an asteroid strike, as well as through various kinds of human action. We have not taken steps toward either one; for nearly half a century, despite the dedicated effort of a small number of astronauts and space advocates, we have done no more than maintain a limited human presence in low orbit. Society as a whole has made no effort at all.

One reader told me he felt that we should spend no more on space travel until every child on Earth is well fed. That dismays me, as I believe the all-too-common idea that we should solve the problems on Earth before moving outward into space is a self-defeating policy. If we wait until we have eradicated poverty to colonize other worlds, neither will ever happen. Expansion into space is the solution—and in my opinion, the only solution—to Earth's problems. Abolishing hunger and pollution and war depends on the use of extraterrestrial resources. The fact that these problems still exist despite well-intentioned efforts to eliminate them is the result of our confinement to a single planet that we have outgrown, and they will inevitably continue to worsen until we make the effort to expand our civilization beyond it.

Most people assume either that we will someday learn to prevent war, or that human nature will eventually lead to interplanetary wars. In my opinion neither of those things will occur. War cannot be abolished by "learning" to prevent it. Negotiation is meaningless because no matter how many leaders negotiate in good faith, there will be fanatics who ignore treaties, and as long as these fanatics can attract enough followers to launch attacks, a strong defense against them is essential; failure to maintain it would lead to worldwide dictatorship. I believe that in time there will be an end to war, but this cannot happen until it becomes impossible for aggressors to recruit a significant number of supporters, a situation that can be brought about only by eliminating the factors that cause people to support them.

Human nature leads to war for two reasons that can't be merely wished away (although the majority attitude toward war certainly becomes more negative as the centuries pass). In the first place, humans crave challenge and excitement, which is the reason our species has been able to survive and thrive; so when a society is not fully occupied with a constructive challenge, it fulfills this need through a destructive one. In the second place, people fight over land and resources when these are scarce or seem likely to become scarce—again, this is an instinct indispensable to survival.

But once we break free of the confines of our native planet, there will be plenty of constructive challenge in the process of surviving elsewhere, and neither living space nor resources will ever be scarce again. The universe contains sufficient resources to last virtually forever—and making use of them needn't involve stealing them from extraterrestrial races; the discovery of numerous exoplanets indicates that there are more than enough uninhabited worlds to go around.

As long we are bound to a single world with shrinking resources, however, the situation can only get worse. There is nothing surprising in the rise of militant groups and terrorists; how could it be otherwise when it's obvious that Earth's resources can't last indefinitely and some societies either have less than others, or fear that what they do have will be taken away? When the frustration of the have-nots, and their lack of any way of constructively changing their situation, makes them easy prey for fanatics who know all too well how to satisfy their instinctive longing for an exciting challenge? This was always true, but in the past trouble wasn't widespread enough to threaten the existence of Earth's civilization as a whole. With modern technology, it becomes increasingly possible for a small minority to endanger the entire planet, even without the use of nuclear weapons (or with them; it no longer takes a superpower to launch a nuclear attack). Yet with that same level of technology we could extend civilization beyond the planet so that even if the worst should happen here, our species will not be wiped out.

This is what the Critical Stage is, and far from being an outdated concept, it becomes more and more pertinent year by year. I see this as a natural stage of evolution. We don't have wars because we are foolish or morally deficient (although there will always be individual evildoers), and we won't have them when our species is mature enough to take up the challenge of interplanetary expansion. I don't believe there will ever be interplanetary war, as many people think is inevitable in view of past history. Conditions will not be the same as in the past. In the terminology of anthropology, war in the past was adaptive for our species—it led step by step to the development of the technology needed to access the resources of a new ecological niche. It will not be adaptive once we are occupying that niche.

Nor will people's attitudes be the same. Centuries ago, war was considered glorious and men felt deprived of opportunity when no war was in progress. Even as recently as World War I, young Americans who joined up were afraid it would be over before they had a chance to get into the fight. Nobody in our culture feels that way today—it's generally agreed that war is a bad thing to be avoided whenever possible. Progress does occur over time. But it takes time—evolution is not a process that can be speeded up by decree, although it can, unfortunately, be stalled by apathy.

For many years I was increasingly worried, not so much by my awareness of more threats as by the fact that nothing was being done to speed up our progress in space and the general public cared less and less about it. I was afraid that our Critical Stage might be unnaturally prolonged. Then, a few years after the republication of *The Far Side of Evil*, it dawned on me that the public's decreasing interest in space is due not to apathy, but to fear—not conscious fear, but the stirring of an unconscious recognition that the universe is very much vaster, and more scary, than most people like to think. (See "Achieving Human Commitment to Space Colonization: Is Fear the Answer?" at my website,) At the time of Columbus, many thought venturesome ships would fall off the edge of the world, a prospect they viewed with great dismay; others (according to legend), knowing the world extended beyond their maps, marked the edges with the warning "Here Be Dragons." Figuratively speaking, most people of our time, having been shown that travel between worlds is no mere fantasy, may feel the same way about space exploration.

And so for a while I thought that the alternative fear of such disasters as biological warfare, environmental deterioration, and terrorism might be the spur needed to get the space program moving again—we wouldn't have gotten to moon without the fear that the Soviets would win the Cold War. But in 2012, while revising my nonfiction book The Planet-Girded Suns for republication, I suddenly saw the striking parallel between today's widespread underlying fear of what the universe may hold and the feeling that prevailed in the seventeenth century when the orderly Earth-centered conception of the cosmos was being replaced by realization that the universe has no center and the stars aren't fixed to a solid crystal sphere. The deep feeling of insecurity this new outlook engendered among the majority of educated people lasted for nearly a hundred years. Is there any reason to assume it will take less time for the public to get used to awareness that humankind is not isolated from whatever exists elsewhere? (See "Confronting the Universe in the Twenty-First Century," published as an Afterword to The Planet-Girded Suns and by The Space Review.)

And so I now think that a commitment to large-scale space efforts will not come soon, and that far from being a sign that something has gone wrong, this is a normal phase of evolution that should have been predictable. The Critical Stage simply isn't as brief as I once believed. That's an optimistic view, as it means we are still on track. But of course the danger of self-destruction remains, and will last until we do take major steps toward space colonization. There is a longer period of peril than I supposed, and thus greater odds that we won't survive it.

In the novel the Service is searching for "the key to the Critical Stage" that might enable them to save other worlds, and when readers asked why I didn't let them find it, I've replied it was because I didn't know the key myself. I now suspect that I do know, and that the only key is time. Thus there is indeed one unreasonable premise underlying the story—the assumption that the Service wasn't already aware of that, considering that it knew the histories of the many worlds in the Federation it represented. However, I naturally don't pretend to portray the very advanced interstellar civilization in my fiction realistically, so I trust that this newly-discovered plot hole is not too serious a flaw. Certainly their immediate concern is valid, since if no start were made toward developing space technology, a world's Critical Stage would end sooner or later in disaster.

Of course there are many individuals in our society who don't share the prevalent uneasiness about human contact with the universe and are enthusiastic about exploring. There will be small-scale activity in space, including bases on the moon and Mars, long before our Critical Stage is over. We are already beginning to make progress with commercial space ventures, which I have always believed are what are needed to bring about significant development of extraterrestrial resources. But I no longer believe we will see any major effort toward colonization before the end of the twenty-first century.

What will happen when our world's Critical Stage finally ends? By definition, we will have begun to spread into space, and more resources will be available to Earth. But I don't think our world will become the utopia many people hope for until much later; I suspect that for the foreseeable future such a society will be possible only in the colonies. I do believe war will be abandoned, yet there will still be troublemakers and police will be needed to deal with them. There will be hunger and poverty because Earth will be overcrowded for a long time to come, and it won't be possible to import sufficient resources until orbiting colonies—as distinguished from those on other planets—are well established (although implementation of space-based solar power could go a long way toward minimizing these problems). And there will be depression and apathy among the majority of citizens who cannot personally participate in the exploration and settling of the frontier. My novels Defender of the Flame and Herald of the Flame, which are set long after many worlds have been colonized, portray what I think is most likely; they paint a dismal picture of conditions on Earth but offer hope from an unexpected direction at the end of the story.

Today's space enthusiasts naturally resist the idea of there being a natural explanation for the slowness of our movement beyond Earth, one that time alone can overcome. My published essay about it wasn't warmly received. Perhaps I am able to believe it only because now that I'm past eighty I know colonies can't be established in my lifetime anyway, nor will I live to see the worsening conditions on Earth likely to prevail before they are. But if my theory about the delay is true, at least we have no present cause to doubt that, barring catastrophe, we will someday reach that pivotal point in our evolution.

Space Colonization, Faith, and Pascal's Wager

This essay, which was published online in The Space Review *on July 3, 2017, was written in response to an article in which the author implied that because the belief that space colonization can ensure the long-term survival of humankind is based on faith, that goal is not realistic enough to be worth espousing.*

*

In his essay "Escaping Earth: Human Spaceflight as Religion" published in the journal *Astropolitics,* historian Roger Launius argues that enthusiasm for space can be viewed as a religion. He focuses mainly on comparisons with the outer trappings of religion, many of which are apt; but in one place he reaches the heart of the issue. "Like those espousing the immortality of the human soul among the world's great religions . . . statements of humanity's salvation through spaceflight are fundamentally statements of faith predicated on no knowledge whatsoever."

I think Launius may be somewhat too pessimistic in his assertion that we have no knowledge whatsoever about our ability to develop technology that will enable humans live in the hostile environment of space, but that is beside the point. It's true that we have no assurance that the colonization of space will ensure the long-term survival of humankind. "Absent the discovery of an Earthlike habitable exoplanet to which humanity might migrate," Launius continues, "his [sic] salvation ideology seems problematic, a statement of faith rather than knowledge or reason." And the accessibility of such an exoplanet is questionable, since according to current theory it will not be possible to cross interstellar space rapidly enough to achieve much migration.

It is indeed faith that underlies the conviction that traveling beyond our home world will prevent the extinction of the human race. But Launius' presentation of this fact seems to imply that it lessens the significance of such a conviction, as if beliefs supported by mere faith were not to be taken seriously. That is far from the case, as the history of human civilization clearly shows.

Most major advances have been made by people who had faith in what they envisioned before they were able to produce evidence; that was what made them keep working toward it. Having faith in the future, whether a personal future or that of one's successors, has always been what inspires human action.

On what grounds can faith without evidence be justified? This issue was addressed by the seventeenth-century philosopher Blaise Pascal in what is known as Pascal's Wager, now considered the first formal use of decision theory. Pascal was considering whether is rational to believe in God, but the principle he formulated has been applied to many other questions. In his words, "Granted that faith cannot be proved, what harm will come to you if you gamble on its truth and it proves false? If you gain, you gain all; if you lose, you lose nothing." If on the other hand, you bet on it being false and it turns out to be true, you lose everything; thus to do so would be stupid if the stakes are high.

This is a clear-cut defense of faith. The many refutations of Pascal's Wager concern not its logic but its unstated premises: that if God exists then there is a Heaven, that only believers go to Heaven, and that belief requires no action leading to harm. Similarly, there are unstated premises in application of this logic to faith in space colonization—it is based on the assumptions that humankind cannot survive indefinitely while confined to one small planet, that extinction of humankind would be a bad outcome, and that widespread faith within society will be needed if large-scale colonies are to be established.

Although many people deny the first assumption, that strikes me as a heads-in-the-sand attitude. It is simply not reasonable to believe that the dwindling resources of Earth, however extended by wise use, can support its population forever; and even if they could, the danger of disaster—whether from global war, runaway technology, or a natural event such as an asteroid strike—would remain. Supposing a remnant of the human race did escape destruction, in the distant future it would be annihilated by changes in the sun.

The second assumption, unlike the first, is not subject to factual analysis. There are people who aren't bothered by the prospect of extinction, and if they're not, no argument can convince them that it is to be avoided. One might ask, however, what difference the time frame makes to them as long as the human race isn't wiped out while they're alive. If it's okay for us to become extinct, why not a mere hundred years from now? Most people, even those who don't care what happens in the distant future, feel strongly that it would be a bad thing for humankind not to outlast their grandchildren. Yet either the fate of our descendants matters, or it doesn't. If it doesn't, why not let our planet's environment deteriorate and save ourselves the trouble of trying to preserve it?

Yes, as Launius says, the belief that long-term survival of humankind can be achieved through spaceflight is based on faith rather than evidence. So is the belief that it would be tragic for humanity not to survive. Faith in something, at least in our existence not being pointless, is essential to functioning as human beings. Almost everyone has a deep, instinctive feeling that our species will continue to exist when we ourselves are gone. In evolutionary terms, this is an adaptive trait. If we lacked it no progress would ever have been made; our ancestors would have achieved nothing that affected those who came after them. We might, in fact, have died out long ago due to circumstances we lacked the technology to deal with.

Does the absence of evidence that colonizing space can save our species make faith in it a religion? Of course it does. Faith in an outcome beyond our present understanding is what religion *is,* not the rituals it involves or the metaphors it employs for the incomprehensible. To those who feel that calling something a religion impugns its relevance to the real world, I would like to say that their definition of religion is too narrow. The essence of religion is its recognition of reality that we cannot explain in terms of facts we now know. The explanations offered by specific religions, whether or not metaphorical, are not its defining aspect. The acknowledgement that we cannot know everything, and must therefore trust that there is a pattern we cannot see, is the universal concept all religions share.

Primitive religions offered metaphors for facts later explained by science. The extent of which the symbols of today's religions are considered metaphorical depends on the individual (this was the case in ancient times, too, as was pointed out by anthropologist Paul Radin in his classic book *Primitive Man as Philosopher*). It may well be that our present conception of interstellar travel is a metaphor for a reality we cannot yet even imagine. But knowing this, knowing that we cannot count on being able to travel rapidly between stars by means of any technology compatible with current theories, we have faith that our descendants will reach them somehow—because the alternative is extinction, and we just can't believe that humankind will become extinct.

Under the principle of Pascal's Wager, reason demands such faith, assuming that unending survival of the human race is an infinite gain, that permanent confinement to Earth will mean extinction—and, of course, that belief in the efficacy of large-scale colonization will lead to the action required to bring it about. It should be added that widespread lack of faith in a reachable goal would result in a great deal of unnecessary hopelessness as conditions on Earth worsen and more and more people take their heads out of the sand. There are already too many, even among young people, who feel hopeless about the future.

However, the consequences of believing in an unreachable goal must also be considered. If we bet against survival and it's impossible, then nothing will happen that was not inevitable. But if we, as a civilization rather than as individuals, have faith that proves vain, will any harm have been done?

Some people will think so. Pascal considered the cost of vain belief negligible since all it meant in regard to the issue he was considering was that he would have gone to church and followed some religious rules unnecessarily. But the demands of faith in space colonization are greater. Large-scale space colonies cannot be established without a very long head start; they will be costly in time, effort, and funds. Some will say the effort and especially the funds would be better spent on improving conditions on Earth. (Never mind that as in the case of the money spent on space so far, funds cut from the space effort would not actually be diverted to causes deemed more worthy.) Even if such people concede that permanent survival of humankind on Earth is impossible, they may feel it would be better to devote ourselves to maximum comfort while we're here than to worry about the fate of our remote descendants.

The issue is complicated by the fact that at present, it doesn't really matter whether individuals have faith or not. We are generations away from the stage at which prevalent opinion will influence the final outcome. The earliest space colonies will be built by people who want to go into space personally and who may or may not be thinking about the long term—although they may well be concerned about the possibility of disaster in the near future. They will not need faith in the ultimate survival of humanity; enthusiasm for their work will be sufficient motivation. It's possible that if these colonies thrive over a long period, they will spearhead the spread of humankind throughout the universe; Earth may not even be involved.

And yet a time may come when the support of society—colonial society if not Earth's—will be needed if the survival of the human race is to be assured. The difficulty of establishing large colonies may prove so great that faith alone can sustain the efforts of our successors. It is not too soon to start encouraging belief in the goal.

In any case, no one today should feel foolish for having faith that space colonization will prove to be humankind's salvation, unlikely though the feasibility of it may now seem. Consideration of Pascal's Wager shows that when no evidence exists, belief is more rational than denial.

Why There Will Never Be an Interplanetary War

Knowledge of humankind's history is important if we are to avoid mistakes of the past, but it's necessary to put past events into perspective by being aware of how things have changed from era to era—and to realize they will go on changing as we evolve.

*

By far the most common theme of science fiction, at least of science fiction directed to the general public, has been interplanetary war or invasion. This isn't surprising, since it provides more scope for exciting stories than less violent themes. It also makes the stakes in a conflict higher than they would be if only individual characters were involved. And it seems a logical extension of human affairs into the future.

But is this really logical? Should we expect that because war has been common throughout human history, it will continue to exist in centuries to come? I don't think so. I believe that the very concept of an interplanetary war is anachronistic, however reasonable it may seem in the light of past experience.

In the first place, humankind progresses. Yes, if a time machine sent people of former centuries into the era of space travel, they might well find an excuse to fight each other. Setting aside the obvious logistical difficulty of launching an attack on a distant planet, they might attempt to conquer it. Conquest was not frowned upon in the past, and many men were eager to participate and thereby gain glory. But today, few if any people consider it glorious to fight in a war. To defend against aggression, when not necessary for personal survival, is seen as a duty—one accepted gladly by some and with extreme reluctance by others. We have outgrown the perception of war as desirable.

But aren't human beings aggressive by nature? Isn't that an innate characteristic of our species? There is no evidence that this is so. We have a built-in drive to overcome obstacles, which is an adaptive trait essential to our survival; but fighting among ourselves is no longer adaptive—on the contrary, it could now lead to extinction. The obstacles are different now, and to overcome them that drive must be channeled into more constructive action. There will always be individual aggressors, but not enough to seize power. Future humans will find challenge enough in preserving—and spreading beyond—our home world.

In the second place, there will be no reason for colonized worlds to come into conflict. Conflicts arise over competition for land or resources. There will be plenty of land on any world fit for colonization, and a viable colony, by definition, will have either sufficient resources to survive or a way to obtain them from moons or asteroids. It would have nothing to gain by attacking another colony even if it had the means to do so. As for far-future colonies with large populations that have established trade, there might be disputes over specific resources—it would not, however, be feasible to go to war over them. To build a fleet of ships capable of that would cost more than any possible return.

Might not one colony, or confederation of colonies, wish to rule another? This analogy with Earth's history is a common fictional scenario, and on the surface it seems plausible; but again, it is based on the notion that people are innately aggressive. An aberrant individual might want to rule, but he could not acquire enough followers, let alone the technology, to pose a threat. Colonists will be too busy developing their own planets to be swayed by a power-seeker's ambition.

There remains the other common scenario, a fight for independence from a mother world. This is the one thing people of the future would be willing to fight for. Yet the time has passed when armed citizens can defeat the kind of force a world such as Earth could employ An uprising, even a politically successful one, is not a war.

*

So wars between humans on different planets are just not going to happen. But there is the far more common, and potentially more serious, question of wars with hostile aliens.

I deplore the endless stream of science fiction about warfare with aliens. In addition to movies, there is a thriving genre of science fiction novels about military strategy and exciting space battles, which is to be applauded for the recent trend toward featuring female commanders but which cannot help but instill a view of alien species as evil. Exposure to a few such stories is harmless but many readers are addicted to them, and I don't think we want generations to grow up feeling that the universe is a hostile, frightening place. And if we do meet aliens in the distant future, we certainly don't want the population of Earth to have been conditioned to expect war with them.

Yet perhaps, as some scientists now warn, there really are aliens that would view us with hostility. Is it likely that they will attack Earth or its colonies? I think not, because there would be no point in their doing so.

It is true that we have no basis for expecting extraterrestrials to be friendly. We are not in a position to know anything at all about extraterrestrials; thus we cannot assume that their psychology is like ours—though personally I believe some qualities of thinking beings are universal. One thing we can be sure of, however, is that any aliens that have starships are rational, for if they were not, they could not have developed advanced technology.

It would not be rational for them to invade an inhabited planet. Extrapolation from Earth's past history is not valid; the common argument that Europeans wiped out the native populations of the Americas is in no way grounds for a fear that aliens will pose a danger to us. It's simply not relevant to the issue. Europeans colonized the Americas to gain land and riches such as gold. It was an instance of the inevitable conflict that arises when humans are competing for the resources of a single planet. But space is full of resources, and any intelligent species with the ability to travel across interstellar space would have long ago developed the ability to utilize them. There is no reason to compete for them, since exoplanets are far more abundant than species technologically capable of establishing colonies.

Similarly, the argument that non-human species on Earth wiped out others competing for the same ecological niche is not relevant, because the ecological niche of space is large enough for an unlimited number of species. There is no need to compete in order to occupy it. In the very long run there will be competition in the sense that species that successfully colonize that niche will survive and multiply, while those that do not will die out when the resources of their home worlds are exhausted. But there won't be competition of the sort that occurs within the confines of a single world.

But may not some planets be more suitable for colonization than others, so that species seek to acquire the best ones? There are several reasons why this won't lead to war. For one, unless the biological characteristics of species that have evolved separately are more similar than seems likely, "best" will not be defined in the same way by two or more civilizations within range of each other. Even if it is, the chances of two being at a close enough technological level to permit war between them are extremely low, considering the length of time required to evolve to that level and the odds against both of them coming across the same world among the millions of planets that exist. As for detecting a civilization that broadcast its existence just at the time when it was vulnerable, even supposing that the older species hadn't matured beyond the point of committing genocide, invasion would be far more costly than colonizing a world not occupied by a potential opponent.

It is often suggested that some alien species may simply be warlike, a natural supposition for those who believe that our own species is innately aggressive. In my opinion, both views are invalid because they are inconsistent with the concept of evolutionary advance. Alien species progress, just as humans do. If they have reached a stage more advanced than our present one—as by definition they will have, if they are capable of interstellar travel—then they have advanced in all ways, which means they have outgrown aggressive impulses toward other peoples. They may be indifferent to us; they may not want to make friends; but they won't try to conquer us. After all, even at our present stage of evolution, we would not choose to invade *their* worlds.

The idea that aliens might invade Earth for its resources goes way back to H. G. Wells' 1897 novel *The War of the Worlds,* in which Martians come here because the resources of Mars are dwindling. At the time it was written people had become aware that Mars is a dry, barren planet (then believed to be far older than Earth), yet they had observed what they thought were canals, leading to speculation that intelligent inhabitants might be attempting to channel their remaining source of water. "Across the gulf of space," Wells wrote, "minds that are to our minds as ours are to those of the beasts that perish, intellects vast and cool and unsympathetic, regarded this earth with envious eyes." This set the precedent that has formed popular impressions of aliens ever since. The vast difference between this solar system's resources and those of countless systems has not been taken into account.

One modern variation of the theme, taken literally by believers in abduction by UFOs, is that alien visitors want genetic material from humans because their own species is too inbred. This is nonsense—the premise of my trilogy *Children of the Star* notwithstanding, a species that could build interstellar ships would surely know enough about genetic engineering to produce any variations it desired. Underlying this scenario, perhaps, is an emotional conviction that humans are genetically superior to other races, even those possessing more advanced space technology.

*

All this supposes that there are alien species somewhere near our evolutionary level who are aware of us. The chances of that may not be large; we may encounter only far younger species that haven't yet developed space flight, or far older ones to whom we seem too primitive to notice. In my opinion the latter cannot happen because I believe what I've said in my novels about truly mature intelligent species not revealing themselves to younger ones. Contrary to the hopes of some SETI enthusiasts, benevolent aliens are not going to teach us how to solve Earth's problems. Stages in a species' maturation cannot be skipped any more than a young child can be instantly turned into an adult. I think that whatever such beings may be observing us, or may observe us in the future, will allow us to evolve naturally and fulfill our own potential without interference.

The premise of my novels is that each separately-evolved thinking species has unique qualities and a unique history, and therefore has something new and revitalizing to bring to a federation of advanced civilizations when it is mature enough to join with them. Premature contact would result in the loss of this unique contribution—young worlds would become no more than copies, probably poor copies, of the older ones. The younger races would be absorbed or they would die out. Even knowing about the existence of a more advanced civilization prior to contact would be harmful, as it would cause them to feel that any effort they might make toward progress was mere duplication and therefore meaningless. And I am convinced that this is true in reality.

Thus the view of aliens expressed in the novels is meant to be taken literally, with one exception: the invaders in *Enchantress from the Stars* are an anachronism. They are no more realistic than the woodcutters who seek to slay dragons, as for literary reasons both were based on mythology of the past or present. And of course the portrayal of the mature civilization is also mythological in the sense that it is based on current conceptions rather than facts beyond our present comprehension.

That some equivalent of the Anthropological Service does exist, I have no doubt. If it did not, young intelligent species could not evolve because they would be influenced by well-meaning older ones, even if no starfaring species are hostile. To be sure, some people view this as a desirable goal. Many, not only believers in channeled spirits or UFOs but some SETI scientists, think that advanced aliens, or their example, could solve all Earth's problems for us if contact were made. There is a long tradition, dating back to the early twentieth century, of assuming that advanced beings would surely step in to relieve human suffering and that their failure to do so means that none exist.

It is rarely held that evolution is a fundamental aspect of the universe, equally applicable to Earth and all other worlds in the cultural as well as the biological sense, which cannot be violated without harm. But that is basis on which I form my opinions about aliens, and why I think that war with advanced ones will never happen. Any starfaring species, including ours, will be far past the stage of having anything to gain by it

,

Humankind's Future in the Cosmos

This essay is a revised version of one written for my 2020 ebook The Future of Being Human. *The original version included a section on space activity in our own solar system, which I have deleted because any speculation about the moon, Mars and subsequent near-term developments will soon be outdated. My thoughts about the distant future, on the other hand, will remain mere opinion until a much later era when knowledge we cannot yet even imagine either supports or invalidates them.*

*

I don't believe life in the future will be as different from today as most current speculation suggests. Artificial intelligence (AI) will alter many aspects of it, but will never take over from humans. Nor are we going to become cyborgs, though we will benefit from genetic and neurotechnological advances where they are useful. No matter how many new technologies are developed, people will have the same underlying wants and needs they do now, as no doubt they did in ancient times. Couples will still fall in love. Their feelings about their families will remain unchanged, as will individuals' hopes and fears regarding what's important to them, however unlike today's their appearance and activities may be.

This will be true, I believe, as far into the future as it is possible to imagine. Human nature isn't going to change. Whatever it is that makes us human—which is at present beyond our understanding—is not subject to transformation by time. But as far as humankind as a whole is concerned, there will be one major step in our evolution that will redefine our status forever. We will become a spacefaring species.

The expansion of our civilization into space is vital if we are to survive indefinitely. There are four major reasons: first, because we are vulnerable to a number of catastrophes, human-caused or natural, as long as we are confined to a single planet. Second, because sooner or later Earth will run of resources, no matter what is done to conserve them; the claim that they can be made sustainable forever is a dangerous illusion. Third, because like all species we have a built-in drive to increase the population, and attempts to frustrate it would lead only to setbacks such as war, pandemics, or mass starvation. And fourth, because throughout human history exploration of new regions has led to renewed creativity and intellectual progress; in time we would decline from boredom if we never moved outward, even if not from the other perils.

In addition to these survival imperatives, we need to become spacefaring because it will bring great benefits to Earth. Initially, the development of space-based solar power will solve the energy crisis; bring about a significant reduction in atmospheric pollution; raise developing nations out of poverty; and provide enough cheap power to desalinate sea water, a process that will be necessary if climate change requires an increase in irrigation, as well as to meet the needs of a growing population. Next, manufacturing in orbit, using raw materials from the moon and asteroids, will result in further reduction of pollution, plus lowered cost of minerals and products that can be imported to Earth. And in the distant future, hunger and poverty will be relieved by relieving overcrowding through population growth on other worlds.

Eventually, settlement of such worlds will reduce the pressures that result from confinement of our species to a single planet with finite space and finite resources. This in turn will lessen conflict and, ultimately, bring an end to war. Once sufficient material resources are available to all nations and a commitment is made to the challenge of establishing large-scale colonies, the reasons for war will disappear, although it will always be necessary to maintain a defense against terrorists and attempted dictatorships. Only expansion beyond Earth can bring this about. If we fail to make the effort then sooner or later, our species will die out.

Assuming that we do expand throughout the solar system, however, eventually we will want to reach the stars. It is not in the nature of humans to stop exploring; even if the resources of our home system prove adequate, the drive that has ensured our survival will still exist. Yet interstellar travel is not merely an extension of space travel as we have known it. With the fastest ships conceivable under the laws of physics as presently understood, it would take hundreds of years to get from star to star. Generations would live and die aboard before their descendants reached the destination, and centuries more would pass before people on Earth heard that they got there. This may be tried, but it would hardly satisfy the need of our species to progress. To do that, it will be necessary to travel faster than light itself.

According to what is known of physics today, faster-than-light travel is absolutely impossible even in theory. I believe a breakthrough will come, but there's no way of predicting how soon; it could be in the twenty-second century or it could be much later. It will necessarily involve some form of what science fiction refers to as a space warp or wormhole, though those are merely metaphors for ideas that are not yet understood.

Or perhaps some principle beyond our present conception will be discovered. It should not be forgotten that progress accelerates. Well over two million years passed from the beginning of the Stone Age to the beginning of the Bronze Age; about 2500 years from then to the era of ancient Greece; roughly 1000 to the start of the Middle Ages; only 300 from the Renaissance to the Industrial Revolution; and a mere 200 more until the 1first moon landing. In the half century since then computers, the Internet and mobile phones have unexpectedly transformed our way of life. So any conjectures made today about the establishment of interstellar colonies are probably far from the mark.

Nevertheless, I feel sure that eventually there will be some means of traveling, or at least communicating, between the stars in a reasonable length of time so that human civilization, and not just humans in the biological sense, can endure and spread throughout our galaxy. Otherwise. the struggle and pain of our evolution—and the creativity—would be pointless, and I have never seen any justification for doubting that these have meaning. All doomsaying of the past has proven wrong. So I will express some thoughts, not as predictions but simply as ideas about what may happen in times to come.

Sooner or later we will have settlements on planets of many stars. Eventually Earth may become a backwater world, honored as our ancestral home but no longer central in human affairs. The human population will be spread among its colonies, which I envision as self-governing and perhaps quite different from each other, with free trade carried on between them. If faster-than-light communication across interstellar distances proves impossible by means of technology, links may be established via telepathy, which is instantaneous; we can't begin to guess what capabilities our descendants will have.

Most people won't have opportunity for interstellar travel, which will surely be expensive; if communication technology proves slow, they may not even be aware of planets other than their own. But they will all be human. I don't think they will evolve into separate species as is sometimes suggested. They will adapt to new environments by means of technology, as humans have always done, not by natural genetic change of a kind that would prevent interbreeding—the latter would require centuries, and technology will surely progress much more quickly even if not adequate to begin with.

Whether technology is used to enable humans to live in new environments or to change environments to meet the needs of humans remains to be seen. Either way, pioneering on exoplanets won't be like the traditional image of farmers taming a fertile new land. The worlds or the people, or both, will have to be modified. We don't know how hard it would be to terraform an alien planet, but it would certainly take too long for settlers to survive during the process without equipment such as breathing masks. We do know that to change humans drastically by genetic engineering would require either that parents raise children very different from themselves, or that embryos be created in laboratories and raised in crèches of some kind; and I suspect that people would not accept either of those alternatives. The family is basic to human life. Without families and homes a settlement could not thrive.

Colonists will probably adopt a combination of strategies—terraforming slowly where feasible, but in the meantime becoming so used to necessary technological aids that they seem natural. Insofar as neurotechnological devices such as implants are helpful, they will be used. Relatively minor genetic changes may be made, but not to the extent of affecting sexual attractiveness. People will adapt to such conditions as low gravity or high temperatures simply putting up with them; children born under such conditions may have some physical characteristics unlike their parents but will not be fundamentally different. Humanity lies in inner feelings, not outward appearance.

If terraforming of a planet proves essential, small settlements of experts to oversee it will precede large-scale colonization. It is often said that the initial population of a colony must be large to avoid genetic damage from inbreeding. This idea arose before anything was known about genetic engineering. Inbreeding in itself is not detrimental; for it to be harmful, genes causing damage to offspring must be inherited from both parents, which inbreeding makes more likely. Surely by the time we can reach worlds of other stars, we will know how to detect and eliminate such mutations through germline modification. We may need to do this in cases of radiation exposure, too, although perhaps long-distance space travelers who want future children will bank eggs and sperm routinely.

It has been seriously suggested that since robots will be better able than humans to make centuries-long voyages and to adapt to alien environments, we may leave interstellar colonization to them and never go ourselves. I cannot imagine anything more pointless. Certainly we will send robot probes to investigate other planets just as we have sent primitive ones to Mars, and self-replicating robots (not human-shaped, but in various forms suited to their tasks) will do the heavy work of building colonies. But that will be only a preliminary to human settlement. A colony, as distinguished from a research station, would be useful only as a place for people to live. If it weren't going to be inhabited by humans, there would be no reason to establish it. The purpose of colonizing other worlds is to ensure the survival of our species, which will require new habitats to accommodate an increasing population. What good could it do to populate the galaxy with robots?

An even crazier version of this scenario envisioned by some is that eventually robots themselves will take the initiative and colonize the entire galaxy, or even other galaxies. Furthermore, they see this as desirable, as if takeover of the universe by superior intelligence we have created were the culmination of our evolution . (It's a potentially genocidal goal, since it makes no allowance for any unknown primitive civilizations that may be developing.) As senseless as this seems to me, I must admit that it is a logical conclusion to draw from the premise that mind consists solely of "intelligence"—and it is thus a reductio ad absurdum with respect to that premise. For more about its fallacy, see my essays "Robots Will Never Replace Humans" and "The Roots of Disbelief in Human Mind Powers."

This false premise also leads to another suggestion made by several distinguished scientists, the notion that robots might prepare distant worlds for light-speed transmission of coded information describing individuals' bodies and then reassemble those individuals, complete with their memories, at the destination. I suppose, since people are used to the transporter in Star Trek, the idea of converting humans to energy patterns does not seem too outlandish, and at least one noted theoretical physicist has predicted that the technology will be invented within 100 years. But while transmission of matter may be achieved, human minds are not wholly material. For the same reason that the commonly-envisioned uploading of people to computers will never happen, electronic transportation of them is inherently impossible; even if they arrived in human form they would be mindless zombies. The idea is simply another consequence of the materialistic assumptions that limit today's science to a narrowly-defined segment of reality.

*

What about contact with extraterrestrial civilizations? According to those who believe robots will colonize the entire cosmos, the fact that they haven't already done so means that there are no civilizations more advanced than ours. I'm more inclined to believe that it means advanced aliens haven't produced such robots because it's not possible. Nevertheless, there is a long tradition of believing that the failure of aliens to show up means there aren't any.

Even before interstellar ships were imagined, at least one well-known writer argued that advanced beings could travel as disembodied spirits and would surely want to end human suffering, so their absence must mean that we are alone. Now some space enthusiasts want to believe we are because they fear that aliens not having come here might mean FTL travel isn't possible. Others believe it because they are convinced that it's our destiny to seed the universe with life, which in my opinion is a revival of the ancient hubristic idea that Earth occupies its center in terms of importance. SETI enthusiasts are beginning to believe it because we haven't received any radio messages. And of course, some people want to believe there are no aliens because they conceive of them as hostile beings who might sooner or later invade Earth.

The absence of aliens is known as the Fermi Paradox because physicist Enrico Fermi long ago expressed surprise that no evidence of their existence has appeared. Personally I have never been able to see anything in the least paradoxical about it. Many plausible reasons why they haven't contacted us have been suggested, but I think the most likely one is that they have chosen not to do so. I don't believe such contact is going to happen—via SETI or in any other way—in this century, and maybe not for many centuries.

As stated in my novels, I believe truly advanced "human" species do not reveal themselves to less advanced ones because they do not want to interfere with their evolution, A speculation similar to this is generally known as the "zoo hypothesis," but that is not the same thing. When we observe animals in zoos, we do not expect them to someday become our equals. We don't think they will progress while we are watching; on the contrary, we take it for granted that their capabilities will remain the same generation after generation. Planetary civilizations, on the other hand, advance. And it seems to me that ETs significantly ahead of us would want us to reach our full potential before joining them, not only for altruistic reasons but because they would value the contributions that diverse civilized species would make to the supercivilization of which they are members—not to mention that they'd realize what trouble admitting an immature species into that confederation might cause.

Long after publication of the novels in which I first

expressed this view, a few writers of scientific papers suggested it, though it is not often taken seriously, no doubt because it eliminates all hope of success in SETI. I have since learned that it was earlier proposed by the Russian rocket pioneer Konstantin Tsiolkovsky; he speculated that advanced beings haven't yet visited us because when we have evolved further we "can bring a new and wonderful stream of life that will renew and supplement their already perfected life." However, he believed that Earth has been isolated only because we are exceptionally promising, and that advanced species do visit other worlds, intentionally interfering with their evolution in order to transform their inhabitants into perfect beings like themselves.

The majority of people who have pondered the distant future, both in earlier eras and in ours, have believed that the ultimate goal is to end all suffering and that sufficiently advanced beings have done this. Many, like transhumanists today, have envisioned the abolishment of death. The Russian cosmists, of whom Tsiolkovsky was one, even believed that all who died in the past could be brought back and that the purpose of interstellar colonization was to make room for them.

That suffering and death are inherent in living seems to be recognized mainly by philosophers who don't envision any future progress other than social reform. Yet denial of this basic fact is contrary to everything we know about life. Some causes of suffering may be eliminated as we evolve further; for example, war, starvation, and prejudice will in time end, and characters in some of my own novels learn to turn off physical pain. But emotional suffering of one kind or another will always exist because it is a fundamental part of being alive. If this were not so, there would be no motive to make the effort living requires, much less to progress.

Some readers of my novels have felt that the advanced civilization I portrayed was rather arbitrary in decreeing that the younger species were not their equals, and certainly I have maintained, in both fiction and nonfiction, that individuals of cultures at all stages are equally human. I extend that principle to people of extraterrestrial species also. But it isn't a matter of innate qualities, for culture has a bearing on the development of minds. The science of epigenetics is discovering that this is true even on the physical level; DNA is not the sole determinant of a person's brain.

Thus cavemen who didn't yet have tools or a spoken language were certainly not our equals; if there were a time machine, we couldn't bring a caveman into our time and expect him to hold his own, even if he was educated. And if we were to hear from an extraterrestrial civilization that has existed for many millennia longer than ours, I don't think we would expect to be the equals of its members. We'd expect them to be advanced in ways we can't even imagine. This natural assumption is obscured in my novels because I have to make the characters enough like us for readers to identify with. I chose to make them physically similar not only for plot purposes but as a literary device to suggest the universality of values and feelings among thinking beings throughout the cosmos.

But if they do exist, they are not so much like us either physically or culturally. And if there is such a thing as progress— which I maintain that there is (in contrast to official scientific theory, which defines evolution simply as "change" without any "forward" movement—then they have developed abilities far beyond ours. Since I could hardly depict advancement in ways we can't imagine in my stories, I tried to symbolize this with the controlled psychic powers. The implication of those powers being common everyday abilities in an advanced civilization may not be apparent—but of course, if people were not hiding them as my characters do on younger worlds, it would be a very different kind of society than has ever existed at any time in human history. Could people of our time get along just fine among beings who habitually communicate telepathically and move things around with psychokinesis? It would be as if a caveman who couldn't learn to speak tried to function among modern Americans.

What's more, the vast majority of human beings as we know them not only are incapable of using controlled psychic powers, but would be all too likely to do harm with them, unintentionally or otherwise; a culture based on them with members no different from ourselves is inconceivable. It goes without saying that I don't know just how a species gets from "here" to "there," to the stage where the widespread use of these powers is both possible and practical—but I believe it happens. In the context of my fiction (which should not be taken as a literal definition of species maturity) this is the turning point. A species is accepted into a supercivilization, if one exists, when it has reached the stage where psychic abilities can be used by its normal members without posing a threat to anyone or finding themselves at a disadvantage. It is not a matter of subjective judgment.

This is only an example, of course. We do not know what the actual turning point is, but it may be something beyond our present ability to envision. But surely development of starships is a prerequisite, as without them we are far behind species who can travel between the stars.

Might we not come into contact with species at our own level or younger, who have not yet decided to conceal themselves? The chances of that are extremely small, considering the vast number of stars in the universe and the vast distances between them. SETI listens for signs of supercivilizations, not isolated ones that can just barely transmit signals and which only by incredible coincidence could be close enough to detect. And when we begin to explore, it will be only by luck that we find even one suitable planet within range, let alone an inhabited one.

It may be wondered why, since we are able to find exoplanets by detecting their effect on observation of stars, we won't eventually detect the presence of a supercivilization's home worlds. This issue was raised by my editor way back in 1969 before *Enchantress from the Stars* was published, before we had found any exoplanets at all. I have always assumed that supercivilizations are able to shield their worlds from detection or intrusion. We don't know of any technology that could accomplish that, but it's no more unlikely than a great many other things we don't know. It would be a useful technology even apart from altruistic aims if a world feared aliens might be hostile.

However, I don't believe there is any need to worry about hostile aliens as some scientists have begun to do, despite the well-established precedent set by science fiction. I have discussed this issue in my essay "Why There Will Never Be an Interplanetary War." In the first place, a starfaring species would have no reason to attack, as the universe is full of planets and resources that any species with the ability to cross interstellar space could obtain more easily than by invading an inhabited world. And in the second place, such species must have advanced far beyond aggressiveness, which I view not as innate but as a stage of immaturity. The invaders in *Enchantress from the Stars* are an anachronism included because the book is intentionally based on both traditional and recent mythology.

So I don't expect us to meet extraterrestrials, hostile or otherwise, in the foreseeable future. If there are any supercivilizations they will conceal their existence even if they are observing Earth, which it's possible that they may already be doing in ways we cannot detect. Surely in the future they will observe our colonization of exoplanets, which may occupy our full attention for centuries.

There will come a time, however, when settling new worlds is not enough. Humans require challenge in order to thrive, and in time colonization will no longer be sufficiently challenging. I believe the next step in evolution will be the acquisition by large numbers of people of the ability to use psi consciously, which will mean radical change on both the individual and the social level. But even that will not permanently satisfy the need to progress into the unknown. And without progress, humankind will inevitably decline.

I see only one way forward: contact with advanced extraterrestrials, when we are at last ready to meet them as equals. We will then confront, as the hero of my novel *Herald of the Flame* says, "a universe larger than the one we've been living in, a universe full of alien worlds with their own people, their own civilizations, a multitude of worlds that will take centuries to learn about and explore. A challenge that will last virtually forever. . . Worlds and peoples different from those we know must exist, for if they don't, there is nowhere to go from here—no hope to inspire future generations. Someday . . . humankind must encounter a new universe to explore, or civilization can only slide further downhill."

This will be the true Singularity, the point past which we can make no predictions about how humanity will change. Both literally and figuratively we will enter territory that is now beyond our comprehension. It may be that no such future will come to pass, but we are better off believing that it will than supposing that we're indistinguishable from robots.

Some may wonder why, since I've never conformed to the usual conventions of the science fiction genre, I chose to write only novels about the future. It was partly that the idea of a universe filled with countless inhabited worlds has always fascinated me, but mainly because I believe that how people in our era think about the future is important. Above all, I want readers to look toward it with hope, in the belief that however difficult our problems, and however slowly our species evolves, humankind will continue to move forward. Our world is one small part of a vast, wonder-filled universe that we will sooner or later encounter. People need to think of it in that light.

If you enjoyed this book you may like T*he Planet-Girded Suns: Our Forebears' Firm Belief in Inhabited Exoplanets*, which contains many more details about the views of extrasolar worlds prevalent in the seventeenth through early twentieth centuries, plus a collection of relevant poetry from that era.

About the Author

Sylvia Engdahl is the author of eleven science fiction novels. Six of them are Young Adult books that are also enjoyed by adults, all of which were originally published by Atheneum and have been republished, in both hardcover and paperback, by different publishers in the twenty-first century. The one for which she is best known, *Enchantress from the Stars* was a Newbery Honor book in 1971, winner of the 1990 Phoenix Award of the Children's Literature Association, and a finalist for the 2002 Book Sense Book of the Year in the Rediscovery category. Her trilogy *Children of the Star* was reissued in a single volume as adult science fiction.

Engdahl's five most recent novels, a duology and a trilogy, are not YA books and are not appropriate for middle-school readers, but will be enjoyed by the many adult fans of her work. In addition, she has issued an updated and expanded edition of her nonfiction book *The Planet-Girded Suns: Our Forebears' Firm Belief in Inhabited Exoplanets* (first published by Atheneum in 1974 with a different subtitle) as well as three ebooks of collected essays.

Between 1957 and 1967 Engdahl was a computer programmer and Computer Systems Specialist for the SAGE Air Defense System. Most recently she has worked as a freelance editor of nonfiction anthologies for high schools. Now retired, she lives in Eugene, Oregon, and welcomes visitors to her website www.sylviaengdahl.com, which contains many of her essays, including those dealing with her long-term advocacy of space colonization.

CURRENTLY AVAILABLE EDITIONS OF SYLVIA ENGDAHL'S BOOKS

Click on the title to see the book description, reviews, excerpts, and purchase links. All are available in inexpensive ebook editions; titles marked pb and/or ab also have paperback and/or audiobook editions

YOUNG ADULT NOVELS
Enchantress from the Stars - pb, ab
The Far Side of Evil - pb, ab
Journey Between Worlds - pb, ab

CHILDREN OF THE STAR
(Trilogy - YA, reissued as adult)
This Star Shall Abide (Book 1, aka *Heritage of the Star)* - pb, ab
Beyond the Tomorrow Mountains (Book 2) - ab
The Doors of the Universe (Book 3) - ab
Children of the Star: The Complete Trilogy (Omnibus) - pb

THE FOUNDERS OF MACLAIRN
(Duology - adult)
Stewards of the Flame (Book 1) - pb, ab
Promise of the Flame (Book 2) - pb, ab

THE CAPTAIN OF *ESTEL*
(Trilogy - adult)
Defender of the Flame (Book 1, aka *Passage to Destiny*) - pb, ab
Herald of the Flame (Book 2, aka *A Ship Named Hope*) - pb, ab
Envoy of the Flame (Book 3, aka *Mission to Earth*) - pb, ab
The Captain of Estel: The Complete Trilogy (Omnibus)

YA ANTHOLOGY
(editor)
Anywhere, Anywhen: Stories of Tomorrow

NONFICTION
*The Planet-Girded Suns: Our Forebears' Firm Belief in
    Inhabited Exoplanets* - pb

COLLECTED ESSAYS
From This Green Earth: Essays on Looking Outward - pb, ab
Selected Essays on *Enchantress from the Stars* and More: A
    Sampler - pb, ab
Reflections on *Enchantress from the Stars* and Other Essays
The Future of Being Human and Other

THE FAMILY FINANCIAL BOOK

The book to help make your financial dreams come tru

The Family Financial Book: A Guide to Understanding Every [
Money Matters and Improving Your Finances is designed to
introduce people to the most basic concepts behind financial
planning and is written in easy to understand language. You wc
find a bunch of confusing terminology that leaves you frustrated a
scratching your head. Instead each chapter is designed to allow y
to take action today and get started toward a bright financial futu
Inside this volume topics covered are: savings, investmentssuch
buying a house and stocks, and planning for the future such
asretirement, prenuptial agreements, and college savings pla
Also included is useful information on choosing attorneys, realtor
advisors and other professionals. The author has spent a lot of ti
attending seminars and reading books on the psychology of finar
success and he wants to share what he has learned. The final
chapter focuses on the winning mind set; how to adopt the right
attitude to become a success. Get a copy of this book and get
started growing wealth for you and your family.

Victor Carrion is a full-time Options and Stock
Investor, and prior to that he worked in the real est
field for over fifteen years as an investor, realtor, an
mortgage broker. In addition to studying Finance i
college he has taken course work, attended
numerous seminars, and read just about every boo
investing, asset protection, finance, the laws of
attraction, the psychology of success and the universal laws
money. He runs an online business and is actively involved in
numerous charity organizations. Currently he is working on the p
two of The Family Financial Book and a book on investing in sto
both due out in 2018. He can be contacted at
victorcarrion77@yahoo.com.

ISBN 9780692048764

900

9 780692 048764

9 798985 853292